AF838001

ROBOTICS FOR ARCHITECTURAL RESTORATION

In the current age of automation, there is a growing opportunity to connect design and manufacturing into a single workflow that harnesses industrial tools such as robots and brings their efficiencies to vulnerable building sites within the conservation of cultural heritage.

Robot-heritage integration is key to accomplishing high-profile conservation efforts that can combine technology with site-specific cultural instances and also address the synthesis of design, time, labor, and material. This book brings together years of original research and provides practical applications, case studies with in-depth analysis, and technical guidance, including drawings and diagrams. Codarin proposes a methodology that can be developed and scaled up to maximize impact on the future of the ever-important fields of cultural heritage and historic preservation.

The first book to comprehensively research the intersection of robotics, digital fabrication, restoration theory, and digital technologies in data collection, this will be of great interest to all students and professionals supporting architectural restoration and conservation, as well as those working in robotics and AI for the architectural and construction industries.

Sara Codarin is an Italian architect and educator at Lawrence Technological University, College of Architecture and Design. Her research spans robotic fabrication, digital craft, computational workflows, and generative AI for world-building. She holds a Master's degree and a PhD from the University of Ferrara, where she investigated methodologies for integrating robotics into architectural restoration. She has exhibited projects at the Biennale Architettura 2025 in Venice, Pratt Institute's School of Architecture in New York, Milan Design Week, and Detroit Month of Design. Sara is a techno-optimist, and she loves robots.

ROBOTICS FOR ARCHITECTURAL RESTORATION

Sara Codarin

Routledge
Taylor & Francis Group
LONDON AND NEW YORK

Designed cover image: Torart

First published 2026
by Routledge
4 Park Square, Milton Park, Abingdon, Oxon OX14 4RN

and by Routledge
605 Third Avenue, New York, NY 10158

Routledge is an imprint of the Taylor & Francis Group, an informa business

© 2026 Sara Codarin

The right of Sara Codarin to be identified as author of this work has been asserted in accordance with sections 77 and 78 of the Copyright, Designs and Patents Act 1988.

All rights reserved. No part of this book may be reprinted or reproduced or utilised in any form or by any electronic, mechanical, or other means, now known or hereafter invented, including photocopying and recording, or in any information storage or retrieval system, without permission in writing from the publishers.

For Product Safety Concerns and Information please contact our EU representative GPSR@taylorandfrancis.com. Taylor & Francis Verlag GmbH, Kaufingerstraße 24, 80331 München, Germany.

Trademark notice: Product or corporate names may be trademarks or registered trademarks, and are used only for identification and explanation without intent to infringe.

British Library Cataloguing-in-Publication Data
A catalogue record for this book is available from the British Library

Library of Congress Cataloging-in-Publication Data
Names: Codarin, Sara author
Title: Robotics for architectural restoration / Sara Codarin.
Description: Abingdon, Oxon : Routledge, 2026. | Includes bibliographical references and index. |
Identifiers: LCCN 2025041156 | ISBN 9781032968315 hardback |
ISBN 9781032968247 paperback | ISBN 9781003590873 ebook
Subjects: LCSH: Architecture and technology | Architecture--Conservation and restoration | Manufacturing processes--Automation | Robotics
Classification: LCC NA2543.T43 C63 2026
LC record available at https://lccn.loc.gov/2025041156

ISBN: 9781032968315 (hbk)
ISBN: 9781032968247 (pbk)
ISBN: 9781003590873 (ebk)

DOI: 10.4324/9781003590873

Typeset in Avenir
by KnowledgeWorks Global Ltd.

To the philosophy, naught but the love of wisdom.

CONTENTS

ACKNOWLEDGMENTS

The author gratefully acknowledges the support of the College of Architecture and Design at Lawrence Technological University in the development of this manuscript. Special thanks are extended to colleagues, staff, students, and research assistants for their help throughout this work. The author also wishes to thank the Department of Architecture at the University of Ferrara, along with mentors and professors there, for providing the opportunity to initiate this line of research during the doctoral years. Sincere gratitude is given to all the contributors who generously shared their research and visual material, which greatly enriched the depth and quality of this book.

Chapter 1

Craft-Oriented Automation

DOI: 10.4324/9781003590873-1

Reframing Conservation

There are currently two dynamic areas of knowledge that are evolving in parallel and rarely intersect: robotic automation in construction and digital strategies for the conservation of architectural heritage. A disciplinary connection between these areas would create a mutual benefit and could change the trajectory of future approaches to restoration efforts. Drawing on concepts from preservation practice and architectural discourse, the study within this book proposes a framework for the integration of advanced tools into the interventions intended for culturally significant buildings, particularly in situations where damage necessitates material repair. Within this continuum, conservation is a practice deeply connected to material science that focused on preventing the decay of physical substances and ensuring their durability over time. Preservation extends this concern by operating as an umbrella term that includes policies, social sciences, and wider multidisciplinary approaches to safeguarding cultural heritage. Restoration builds on both, functioning as a critical, design-oriented process that engages with material authenticity and cultural meaning, seeking to reestablish the architectural presence of a work while maintaining its historical continuity.

Industrial robots serve as an entry point for this discussion, not as the sole solution for process development, but rather as a provocation to challenge assumptions and open a broader conversation about construction roles. Industrial robotics arms are machines that carry out precise and adaptable tasks; they interact with different materials through physical attachments capable of cutting, slicing, scoring, bending, extruding, melting, pushing, pulling, poking, and reshaping surfaces and volumes. In this context, robots serve as a metaphor for describing processes that are flexible, responsive, and rooted in direct physical engagement with their surrounding matter. Robots also function as a cultural reference. These automated entities, which resemble human biomechanics, appear widely in popular media—films, graphic novels, animations—as symbol of industrial logic, and they are often associated with the serial production of identical objects. This image stands in stark contrast to the idea of architecture and conservation, because these are disciplines grounded in humanism, where built artifacts are treated as a unique response to site conditions, traditional building techniques, economic needs, and additional contextual factors.

Architectural design is typically understood as a field where repetition is not the norm and where the building process is shaped by variables that resist standardization and encourage individuality. Within this framework, "Robotics for Architectural Restoration" can sound disruptive, or for some, outrageous. The idea of robots sharing a creative workspace alongside humans, even as a hypothetical scenario, challenges the long-standing assumptions we hold about authorship, craftsmanship, and culture. The theory of restoration itself has historically reinforced these beliefs, by institutionalizing the way conservation is practiced, prioritizing transparency and caution, and favoring the more conservative end of the regulatory spectrum. When a historic artifact needs to be retouched, conservation professionals are expected to "confess" each alteration to historians (Otero-Pailos et al. 2016), document every intervention, and make any changes clearly visible to avoid accusations of falsifying the past or "inventing" a new tradition, an artifice in itself (Hobsbawm and Ranger 2012). Some might argue, instead, that any intervention built upon an existing cultural artifact eventually becomes part of its historical lineage, absorbing the principle of haecceity, the quality of being a unique entity rooted in its own essence.

Dissonant scenarios prompt a reflection on how older knowledge, such as manual expertise and the cognitive dimension, can coexist with newly emerging technologies. Their collaboration could reshape and improve contemporary practices. Instead of positioning one approach as superior to the other, the focus would shift toward understanding how each can inform and expand the other to encourage a critical examination of methods, aesthetics, and intentions existing within both fields. To support this line of thought, the book examines relevant case studies drawn from restoration and robotics. Some explore architectural interventions that integrate contemporary additions within the palimpsest[1] of historic structures, while others investigate robotic fabrication projects that demonstrate variation, precision, and customization within their physical production. These examples lead to a hands-on experiment, serving as a test that generated a physical artifact. The result was not a polished outcome, but a low-resolution, low-cost proof-of-concept prototype that supported a legible and tangible process. This phase brought together the discourses of conservation and robotics, envisioning the creative application of advanced workflows in restoration. Engaging directly with tools was necessary to lend credibility within the writing of this manuscript. Unexpectedly, "pushing buttons" in the lab revealed some of the underlying principles of working with robotics, exemplifying that the process depends on the definition of a choreography or a sequence of invisible points in space that the robot's attachment follows to apply force to a material. This spatial organization of machinery can seem counterintuitive to designers, who are typically focused on defining a design vision, much like composing a painting, rather than on the high-fidelity sequence of actions that bring it into being, like the arc of the arm that pushes the brush against paper.

If we understand digital restoration as the act of using digital tools and techniques to repair, enhance, or recreate damaged or deteriorated artifacts, then two complementary references are identified as guiding an overarching narrative: the notions of reliable conservation and experimental preservation. Reliable conservation, a concept developed by Maria Luisa Germanà, considers both material reliability (the quality of operations) and immaterial reliability (the integrity of the process) as essential. These dimensions must be paired with an ethic of responsibility, which Germanà identifies as the foundation for any meaningful engagement with technology (Germanà 2004). On the other hand, experimental preservation, as proposed by Jorge Otero-Pailos, views preservation not as the post-factum protection of culture, but as a form of production in its own right. This view is oriented toward the exploration of new ways of preserving that are practical, forward-looking, and capable of moving beyond institutionalized modes of practice (Otero-Pailos et al. 2016).

The convergence of these two positions opens possibilities and highlights contradictions in examining the broader implications for designers, especially in terms of agency and impact within professional architectural work. The book engages with some of these themes by looking into the relationship between disciplinary needs and technological responses. The methodology that takes shape through this process leans toward a techno-optimistic view, where tools are seen as extensions of physical and design-thinking human capacity. At the same time, the approach resists a technocratic[2] stance. Technology is not assumed to have a force of control, but instead is seen as being a responsive collaborator open to adaptation, adjustment, and reinterpretation as conditions evolve.

Contextual Background

The preservation and recovery of cultural heritage buildings can benefit from the transformative opportunities emerging through process and product innovations associated with the Fourth Industrial Revolution (Schwab 2017). Design and operational workflows are constantly being redefined alongside the rapid development of artificial intelligence (AI), while terminology related to the Fourth Industrial Revolution, such as "Industry 4.0," "Second Machine Age," and "Industrial Renaissance," have become increasingly interchangeable, differing only in subtle nuances.[3] This progressive ecosystem entails methodological and operational advances that have repercussions in the decision-making phases of the architectural project and introduces an era of new and largely unexplored possibilities for environmental and technological design across scales. Although we still reside within the Fourth Industrial Revolution, the concept of the Fifth Industrial Revolution (Sharma 2023) is beginning to take shape and is gaining both traction and attention. This next phase is envisioned as a future phase that builds upon recent advancements and focuses on human-machine collaboration, safety, and the ethical considerations that result of new territories. While exhilarating possibilities lie ahead, the current stage of innovation keeps us occupied with procedural possibilities that are investigated from the construction unit to the technological system. Multiple efforts are underway to integrate computation with digital manufacturing processes into unified workflows. Within this context, computation defines a structured "how-to" process (Abelson et al. 1996), where repetitive procedures enable variation through the adjustment of parameters.

Technological Shift and Need for a Multidisciplinary Approach

In architecture, technological transfers from multiple disciplines increasingly integrate diverse skillsets. Architectural design cannot be regarded as a standalone field within this broader system, where expertise from various domains converges and collaborates (Fig. 1.1). If we frame ourselves in geological terms, within the context of the carbon-heavy economy of industrial age referred to as Anthropocene (Hackett et al. 2018), evolving technologies are shaping an increasingly interconnected ecosystem—one that contributes to the formation of an artificial global layer known as the Technosphere (Alleyne 2019). Within this ecosystem, a systemic evolution of productive culture is driving advances in big data, augmented reality (AR), AI, machine learning (ML), social networks, analytics, and cloud

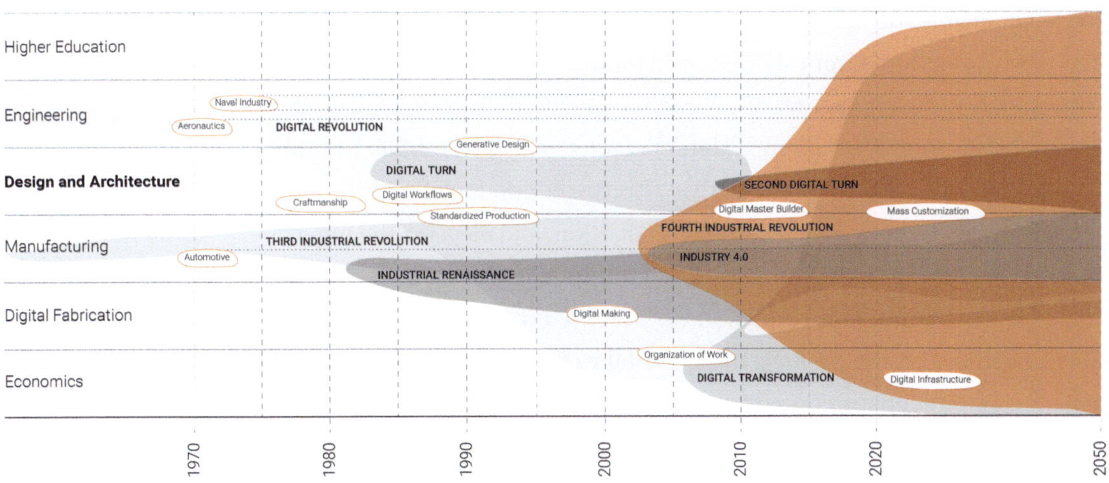

Figure 1.1 Technological framework at large. The Fourth Industrial Revolution is presented as a socio-technical transformation driven by advancements in manufacturing, creating new connections across diverse fields of knowledge and expertise.

storage, hallmarks of a post-work and post-digital society. Compared to previous socio-technical shifts, the pace of the Fourth Industrial Revolution is much faster, marked by ubiquity and intercommunication across production systems.

Focusing on the architectural sector, the anthropogenic marks of the Fourth Industrial Revolution have been theorized by Mario Carpo as the Second Digital Turn (Carpo 2017), characterized by the introduction of automated tools for customized production within the design process. This second phase assumes the existence of an earlier shift, the First Digital Revolution or Digital Turn, which began in the 1990s, when architecture first adopted digital workflows from the naval and aeronautic industries (Carpo 2013). This first transformation introduced a new design culture that advanced computational thinking and digital tectonics, allowing architects to engage with a shared digital language, reduce uncertainty, and make more informed decisions throughout the design process. The First Digital Revolution was marked by a formal elaboration of the digital environment through two-dimensional simulations, placing material culture in a secondary position within architectural discourse. In contrast, the Second Digital Turn is defined by the use of programmable tools that enable the materialization of digital space with greater flexibility and fewer constraints from standardized production methods. These new tools support the exploration of what Neil Gershenfeld described as the Digital Fabrication Revolution, a shift characterized by the ability to "turn data into things" (Gershenfeld 2012). This revolution is defined through digital exchanges between design and production, and elevates the impact of material culture. Consequently, design output is able to link conceptual phases and the built results, as only tacit craft traditions have been able to achieve in the past. With digital fabrication, design is no longer separated from construction, and the translation from one to the other becomes nearly instantaneous (Daubmann 2004), rendering digital space both tangible and perceivable.

This resulting conceptual and spatial continuity creates an opportunity to reintroduce the figure of the master builder as a contemporary expression of digital complexity and dexterity that transcends individual authorship. Such a shift encourages pioneering conceptual outcomes and renewed aesthetic paradigms that redefine the roles and disciplines of professionals collaborating within the digital environment. Digital manufacturing technologies such as robotic arms, 3D printers, smart assembly systems, and hybrid tools play a foundational role in this context. These technologies enable mass customization and support performative approaches to architecture, addressing the shortcomings of the First Digital Turn, which failed to fully cultivate aesthetic and material sensibilities in architectural production (Figliola 2017), and instead pointing toward the emergence of a new digital craft—a craft-oriented automation—open to new multidisciplinary associations, meaning, and interpretations.

Digital Heritage and Demand of Customized Production

Collective architectural heritage is continuously at risk of damage, destruction, or loss due to causes that range from neglect and abandonment to political conflict, environmental degradation, natural disasters, and vandalism. The scientific community has organized knowledge on this topic through collaborative research programs aimed, where possible, at preservation efforts. The *UNESCO*[4] *List of World Heritage in Danger*[5] serves as a key reference document, outlining the definitions of risk and damage conditions affecting heritage sites. This resource provides information on the geographical and socio-economic context of each site, along with assessments of their existing state, maintenance status, and the priority actions required for their protection. As an actionable guideline, the European Commission has issued a document titled *The Future of Europe's Past*, reflecting on a coordinated effort to align member state strategies in accelerating the digitization of cultural heritage in response to contemporary preservation challenges (European Commission 2024). Member states are encouraged to intensify their efforts, particularly by expanding three-dimensional documentation of cultural assets and by adopting advanced technologies such as AI and virtual reality. Historically, building construction has been a conservative field, slow to change and rooted in traditional methods and narrow technological choices. While the industry continues to manage risk through economic, labor, and safety considerations, emerging technologies provide opportunities to redefine construction roles. Digitally driven strategies can address increasing demands for higher quality and faster delivery, even in complex environments requiring attention to environmental, social, and economic sustainability. In the current age of automation, deploying intelligent industrial robots to recovery sites can be envisioned as a viable option. The protection of the site and its elements would entail a customization of on-site operations in a responsive way, respecting boundary conditions and the compartmentalization of automated tools.

The use of digital tools in conservation has largely been limited to surveying and diagnostics, primarily aimed at supporting data capture and digital 3D modeling for programmatic restoration operations. However, the recovery of existing buildings presents opportunities for optimization through the application of contemporary production technologies capable of streamlining the complexity of manufacturing customized architectural elements. Recovery sites, in particular, often require case-specific problem solving and on-site decision making due to the geometric uniqueness of historic buildings. These tailored design solutions must respond to unpredictable variables, including the lack of accurate information about building geometry, unknown assembly methods, and potential structural instabilities. Developing a workflow in this context serves as an update to conventional technical phases, bringing together technological methods and cultural considerations to achieve meaningful design outcomes. This perspective calls for a renewed vision of the master builder, understood not as a singular figure but as a team of coordinated professionals responsible for supervising design, construction, and management activities. Such a collective approach repositions the role of construction workers and offers a possible balance between technological advancement and artisanal knowledge within a complex building process. Years of positive progress in industry and research, especially in computer-aided design and computer-aided manufacturing (CAD/CAM), have established a basis for evaluating successful experiments in digitally informed architecture, creating a foundation upon which digital technologies and fabrication tools can now support the decision-making phases of innovative and context-sensitive interventions in cultural heritage.

Definitions

This research brings together two distinct fields, each with its own strengths and limitations: cultural heritage restoration and robotic construction. To establish common ground, key definitions and recurring concepts are outlined to clarify the objectives and support feasibility assessment.

Cultural Heritage

Historical sources indicate that, as early as ancient Rome, there was an awareness of the cultural value of human-made works. In the 6th century AD, Theodoric the Great, King of the Ostrogoths, referred to the Seven Wonders of the World in his writings, encouraging the Romans to create monuments that would surpass the fame of those from antiquity. This encouragement reflected a broader desire to preserve the scientific and technological achievements of the time, ensuring that future generations could experience them as expressions of identity embedded in materials and civilizations. Over the centuries, architectural monuments have been studied by scholars across the globe. However, a unified definition of heritage has never fully emerged. The restoration charters of the 20th century have helped structure the concept, articulating the need to preserve material objects of cultural value. Cultural heritage was defined and formally expressed by UNESCO in the *Convention Concerning the Protection of the World Cultural and Natural Heritage of 1972*. The document refers to monuments, groups of buildings, and sites as bearing common interest, outstanding importance, and universal value. Heritage is presented as a collective responsibility that concerns all of humanity, independent of time or place. Monuments consist of architectural works, monumental sculptures and paintings, archaeological elements or structures, inscriptions, cave dwellings, and combinations of these features. Groups of buildings are described as separate or connected clusters that are unified through architectural homogeneity or their role in the landscape. Cultural heritage sites refer to works created by humans or through the interaction between humans and nature and are considered to hold outstanding universal value from historical, aesthetic, ethnological, or anthropological perspectives. An additional specification gained prominence in the early 2000s, introducing a distinction between tangible (physical) and intangible (non-physical) forms of cultural heritage (UNESCO 2004). The latter refers to traditions, knowledge, and skills passed down through generations.

Restoration

The most actionable and enduring definition of restoration in practice derives from Cesare Brandi's *Theory of Restoration*, which defines it as "the methodological moment of recognizing a work of art, in its physical being and its dual aesthetic and historical nature, with a view to its transmission to the future" (Brandi 1963). Restoration, in this sense, is a modern construct, emerging alongside a collective awareness of the difference between past and present, a critical distance that enables the redefinition of the "ancient" by framing it within a historical dimension (Torsello 2005).

Closely related to restoration is the notion of conservation. Although the two terms are often used interchangeably,[6] yet they refer to distinct concepts with different aims and conceptual frameworks, both of which concern the care of cultural artifacts and historic buildings. Conservation is concerned primarily with preserving a work in its current state, without modifying its form or function and minimizing deterioration. Restoration, on the other hand, is a critical and design-oriented process, an active reinterpretation of the past that might include integrations, reinsertions, and a conscious dialogue with historical material. As Giovanni Carbonara explains, restoration goes beyond conservative techniques, engaging cultural, aesthetic, and temporal judgments. Its purpose is not only to protect, but also to reactivate the historical and artistic value of the artifact in a way that respects its identity while positioning it meaningfully within the contemporary context (Carbonara 2013). This distinction raises a broader disciplinary question: does restoration define the boundaries of a specialized field, or is it a way of thinking about architecture itself, a mode of action that brings together multiple forms of knowledge and cultural investment? For Manfredo Tafuri, there is no doubt: restoration is not a discipline in its own right, but a subset of architecture; restoration constitutes a transformation that acknowledges historical values, yet remains distinct from conservation. Purely conservative interventions, in his view, are highly specialized tasks best entrusted to technical experts[7] (Tafuri 1991). There are numerous theoretical approaches to restoration—scientific restoration, conservative restoration, post-critical and textual restoration—but they all share a common aim: the conservation of culture (Dalla Negra 2017). In this manuscript the terms restoration and conservation are used with this same intention, grounded in the motivation to conserve culture.

Translating these frameworks into practice, the restoration of buildings with historical value requires careful attention to both aesthetic and historical dimensions, supported by diagnostic surveys and analyses of material degradation. Based on these findings, a coordinated set of actions can be formulated to safeguard these tangible, anthropogenic artifacts of collective value. Such actions are guided by the principles of minimum intervention, reversibility, and recognizability, with the aim of reconstructing the artifact's potential unity without erasing the visible traces of its layered history. Central to this approach is the commitment to preserving as much of the original material as possible, ensuring that the physical substance of the past remains legible in the present. Yet preservation does not always mean retention in a literal sense: even demolition, in specific contexts, can align with conservation objectives, particularly when it facilitates the refunctionalization of a historic building.[8] In this light, leaving a structure vacant is not necessarily the most effective strategy for its protection. Instead, understanding its latent potential, its capacity to take on new functions while maintaining its cultural significance, can offer a more sustainable path toward long-term care and continued relevance. To be effective, restoration must be approached through a critical lens, informed by material conditions, historical inquiry, and a disciplined design culture. Complementary to these are preventive care and research activities that deepen the understanding of an artifact's history and meaning in order to ensure its lasting integrity within the built environment.

Robots

The definition of a robot has evolved significantly since Leonardo da Vinci's 1497 sketch of a mechanical knight. The concept of automatic machines developed gradually over the centuries, particularly through the visionary work of physicists and mathematicians. This growing body of knowledge eventually led to the development of multi-axis anthropomorphic robots, grounded in the patent for the *Programmed Article Transfer* device filed by George Devol in 1954.[9] The International Federation of Robotics (IFR) adopts the definition provided by the International Organization for Standardization in ISO 8373:2021: Robotics—Vocabulary. According to this standard, robotics is defined as the science and practice of designing, manufacturing, and applying robots. A robot is described as a programmed mechanism with a degree of autonomy capable of performing locomotion, manipulation, or positioning. Robots can take various forms and serve diverse functions. This study focuses specifically on tools inspired by human biomechanics, which share the same anatomical terminology used for body parts, such as arm, elbow, and wrist, commonly referred to as industrial robots. An industrial robot is an automatically controlled, reprogrammable, and multipurpose manipulator that operates in three or more axes. It may be fixed or mobile and is used in industrial automation to move materials, parts, tools, or specialized devices through variable programmed motions to perform a range of tasks. The components of an industrial robot typically include a manipulator, an actuator or end-effector, a controller, a teach pendant, and any associated software or hardware communication interface. Over the decades, industrial robotic arms have evolved from single-task, repetitive machines executing predefined sequences into sensor-driven, communicative, and collaborative systems. Their early applications were in the automotive sector, where they were used for mass production. Since then, they have been adopted in multiple adjacent areas of knowledge including architecture and design, where they function as collaborators capable of performing complex, bespoke tasks that enrich design implementation, expand the potential of material expression, and produce new forms of knowledge (Picon 2014).

Digitalization of Design Processes

The digitalization of the design process supports a progression from concept initiation to the production of project documentation in a digitally driven way. This mode of operation is characterized by open-ended and unpredictable, yet consistent, transformations of three-dimensional digital models that generate new architectural possibilities (Kolarevic 2004). The fluid nature of the process has been described as the digital continuum, referring to the digital model used as an interface and communication medium among professionals involved in the project (Leach 2002). As mentioned, the digital revolution of the 1990s brought process innovation into architecture, thanks in part to graphic interfaces and increasingly affordable computing power. This period marked the transition from mechanical to digital technologies. Architecture looked to transportation design as a reference, where boats, planes, and cars are fully developed and tested in digital environments. These industries no longer relied on drawings to construct high-tech, precision products, but instead use comprehensive three-dimensional models that guide the entire process from design to production. The spread of digital workflows in professional practice has allowed architects, even if only in rare, high-profile instances, to use computational data to directly inform building production, where the implications of digitization impact the translation between CAD and CAM. Especially with the adoption of digital fabrication tools, a methodological shift occurs: the 3D model is no longer just a medium of communication but becomes a kind of "shop model" that sends data-rich instructions directly to fabrication. In this context, the model becomes the "thing itself," rather than a "digital representation of the thing." This reorientation challenges the relevance of static 2D drawings during construction, as the process is increasingly defined by data and machine toolpaths. The obsolescence of traditional drawing sets becomes evident as the abstraction offered by 2D drawings is no longer necessary to comprehend the design outcome, which opens the possibility to continuously refine the design process and fully leverage the potential of digitization.

Industrialization of Building Products

In the construction industry, industrialization refers to the delocalized production of standardized technological elements for their repetitive and efficient fabrication. The advent of industrialization in the building sector weakened the link between design and construction by reducing the role of artisanal interpretation (Spadolini 1981) and imposing limitations shaped by production technologies. As a result, architects had to adapt their formal and constructive decisions to align with the constraints of industrial processes. This shift also redefined the role of the contemporary architect, who increasingly operates within the logic of pre-manufactured, off-the-shelf components (Groak 2002). In the 1980s, in a period of technical experimentation in Europe, Pierluigi Spadolini studied and organized construction elements into a systematic framework, dividing the production of the technological units into three categories consisting of industrial, semi-industrial, and craft products. Industrial products are manufactured with identical, fixed features. Semi-industrial products allow for some variations, though these are still constrained by the limitations of the production technology and assembly methods. Craft products are created on-site and are contingent upon manual labor capacity and specific material limitations. Artisanal construction positions the designer as a central figure for the development of all phases. These distinctions parallel two major subsets of construction operations. Closed prefabrication, dominant in the 1950s and 1960s, prioritized rigid, standardized mass production and de-skilled labor, often at the expense of architectural quality. By contrast, the 1980s saw the rise of open prefabrication, characterized by flexible and experimental systems that introduced variation in materials and greater design freedom. This shift constituted a productive context for mass customization, reconciling differentiated production with the efficiencies of industrialized construction. In this way, the idea of a production process including differentiated elements was introduced as exhibiting the same economic efficiency as those produced through standardized processes (Davis 1997). The Fourth Industrial Revolution and the spread of automated systems have opened scenarios for customized on-site and off-site production and offered possibilities beyond the constraints typically imposed by traditional manufacturing methods. These advancements support open prefabrication and mass customization, enabling differentiated responses to complex design inputs.

Investigative Motivations

With the Fourth Industrial Revolution, technology is advancing rapidly and becoming increasingly accessible through online networks. New ideas and products can spread at speeds never seen before historically. In contrast, as the existing workforce ages and retires, the construction industry faces uncertainty, especially as younger generations pursue different career paths. Within this transition, first-world economies are evolving into knowledge economies, where value is generated through expertise, innovation, and intellectual control rather than manual labor. In this context, the master builder re-emerges as a central figure, a professional who demonstrates a deep understanding of processes and instructs tools on how to operate (Fig. 1.2). Designers must recognize these shifts to

Fourth Industrial Revolution	Craft Traditions	Customization	Labor
interconnected tools virtual/physical connections need for advanced computing and scripting skills	workforce will retire skilled workers are expensive current organization of work will be obsolete within a decade	demand for better quality and scope demand for bespoke products demand for data-driven performance	human-machine collaboration machines are tranied to perform problem-specific tasks experience is enhanced by AI

DESIGN PROBLEM FOR THE MASTER BUILDER
understands labor, technology, and culture

DESIGN AND CULTURAL HERITAGE
very unique design problem

Figure 1.2 Design challenges at the intersection of the Fourth Industrial Revolution, artisanal traditions, customization, shifts in labor, digital skills, and human-machine collaboration.

pioneer new workflows and integrated systems for architecture, even as the construction industry remains slow to adopt change. This moment marks an inflection point (Brynjolfsson and McAfee 2014) where digital technologies are poised to reshape production through automation and unprecedented modes of making. Designers are uniquely positioned to lead this transformation by expanding their expertise across computational design, digital fabrication, scripting, and materials crafting in flexible and responsive ways.

Rapid technological development is not always readily accepted by society or by organizations that are hesitant to embrace change. This acceleration is often described by Moore's Law, a prediction made in 1965 by Intel co-founder Gordon Moore, which highlights the exponential pace of technological advancement. To visualize the disparity between technological evolution and organizational adaptation, Martec's Law—a term coined by Scott Brinker that blends "marketing" and "technology," offers an interpretation in a change/time equation: meaning that while technology advances exponentially, organizations evolve logarithmically, at a much slower pace (Fig. 1.3). If the broader building industry is slow to adapt, the practices of conservation lag even further behind due to their dependence on skilled manual labor. An increasing divide is emerging between the tools we use and the cultural context of their application. This disconnect is a key motivation to advance research and develop workflows with the greatest potential for future impact.

Trends in technological development can also be viewed through the vantage point of scientific progress, where older systems are gradually replaced by newer ones. Foster's Curve, attributed to Richard Foster, a former consultant at McKinsey & Company, also known as the "S" curve of technology, illustrates the relationship between time and performance, mapping a technology's transition from early adoption to maturity (Fig. 1.4). Typically, following its initial market release, a technology experiences rapid performance gains alongside decreasing cost. Once it approaches its scientific or practical limits, a new technology emerges to take its place (Foster 1986).

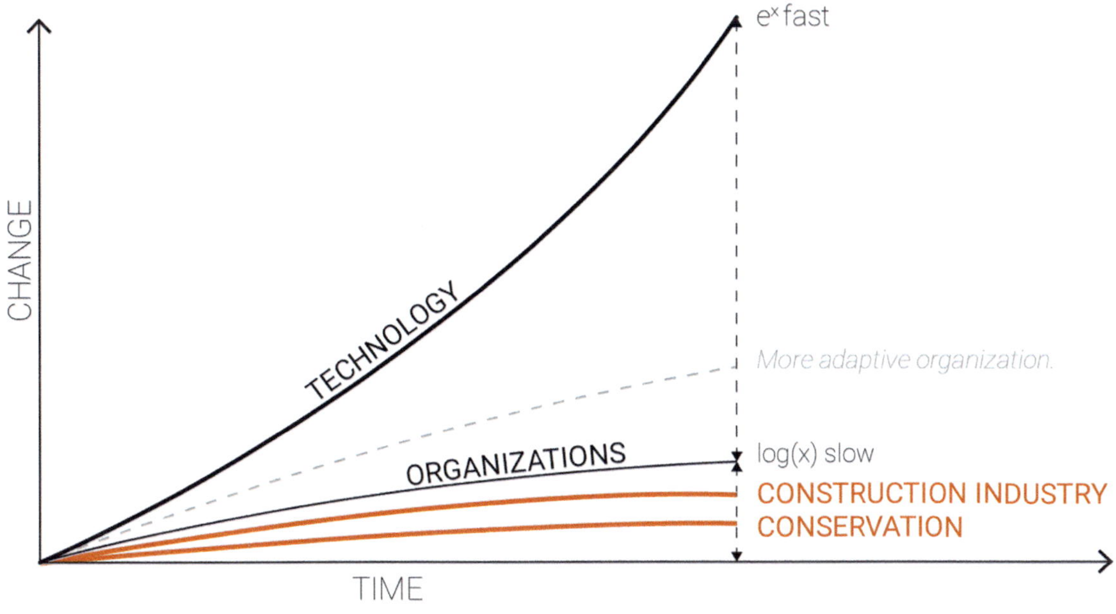

Figure 1.3 Martec's Law reframed. Technology advances exponentially, while organizational change follows a much slower logarithmic pace. Architecture conservation is more resistant to change than the broader construction industry.

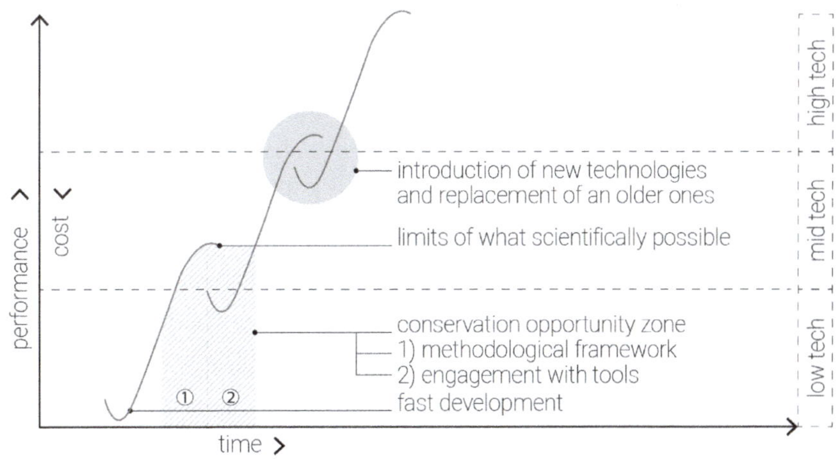

Figure 1.4 The "S" curve of innovation, or Foster's Curve, reinterpreted through the lens of conservation methodologies. The model illustrates how technological performance evolves over time in relation to effort, investment, and development cycles.

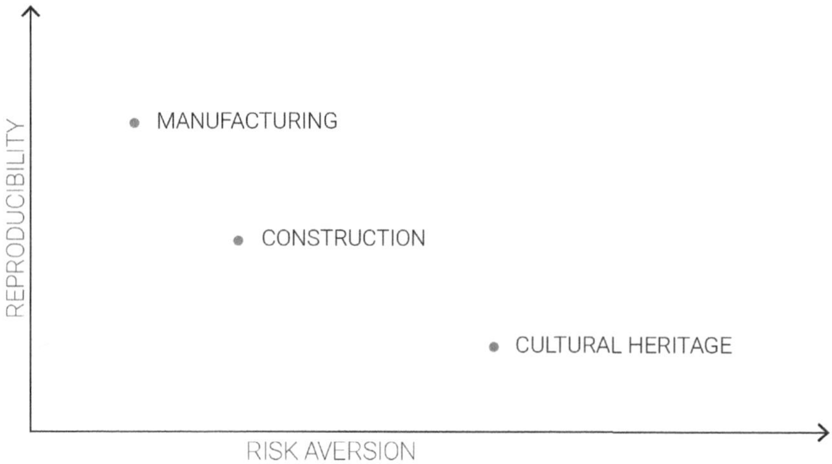

Figure 1.5 The relationship between reproducibility and risk aversion in production processes.

The construction industry is likely to resist change, largely because implementing new technologies requires widespread infrastructural adaptation. In the context of cultural heritage, restoration efforts, particularly those involving monumental architecture, demand careful risk management (Fig. 1.5). The cultural mindset should support the gradual evolution of the discipline and anticipate change when major technological shifts become necessary.

In terms of performance trends, consumers and clients increasingly expect higher construction quality in less time and at lower cost. The traditional linear relationship, where quality (Q) multiplied by scope (S) equals cost (C) multiplied by time (T), has shifted. What was once a balanced equation has become parabolic: $Q \times S > C \times T$. This shift is largely due to the integration of standardized prefabrication processes (Kieran and Timberlake 2004). Rethinking such long-held assumptions may help designers recognize that an exponential increase in profit or revenue is possible through automation and the adoption of new technologies (Fig. 1.6).

Research Methodology and Tools

The investigation adopts an inductive method, progressing from a series of unique cases toward broader explanatory principles (Cross 1984). Observation of qualitative phenomena forms the foundation of the research

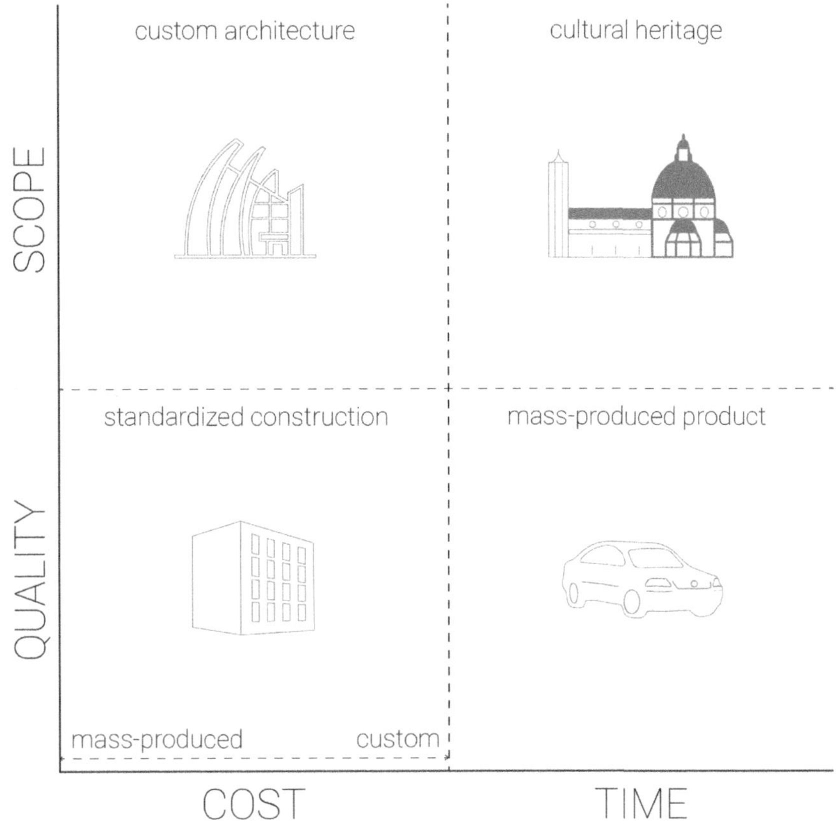

Figure 1.6
Reinterpreted diagram based on Kieran and Timberlake's *Refabricating Architecture*, highlighting how certain design products are more sensitive to variations in specific parameters within the quality × scope > cost × time production equation.

approach (Creswell and Creswell 2017), establishing an analytical trajectory that leads further into data and suggests additional relationships (Yin 2017). The research is structured around four work phases, referred to as Work Packages (WPs), which are sometimes sequential and at other times overlapping (Fig. 1.7). As in architectural research more broadly, this process reflects a "systematic inquiry directed toward the creation of knowledge" (Snyder 1984).

WP1 is dedicated to the acquisition of preliminary information and the definition of the state of the art. This phase focuses on the creation of a bibliography (Fig. 1.8) and the establishment of a structure for the argument (Booth et al. 2003), articulating the rationale behind the research objectives while incorporating feedback from experts consulted during

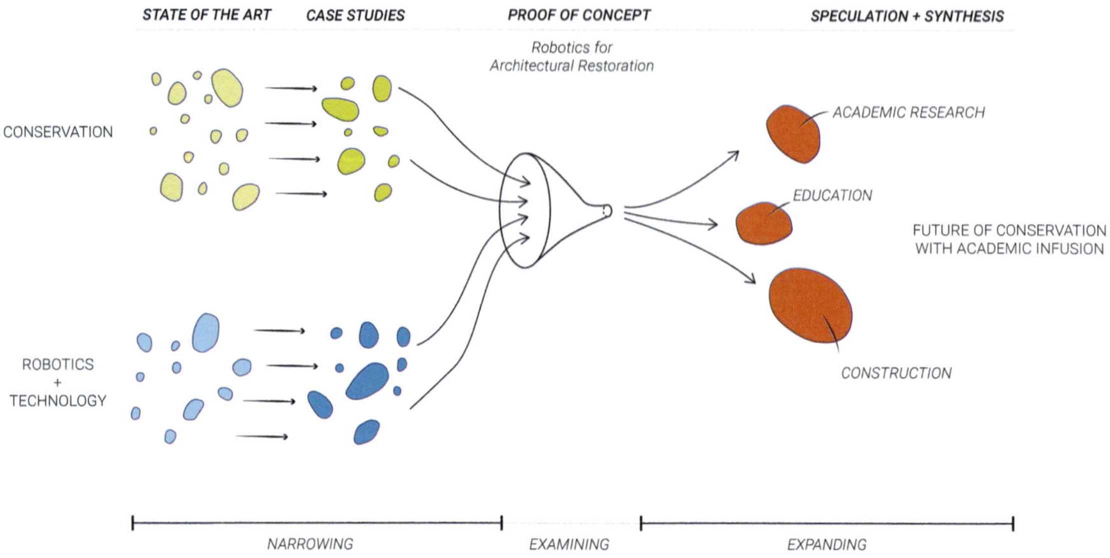

Figure 1.7 Research methodology. The work follows a case study analysis approach, culminating in an applied application to support the subject under investigation. The evaluation of results enabled a broader discussion and facilitated engagement across multiple fields of knowledge.

Figure 1.8 Bibliographic mapping of the research references, organized by publication, year, and thematic focus. Bubble size indicates the density of references per category, distinguishing between technology-oriented sources and cultural heritage-oriented sources.

the process. WP2 consists of data collection through the analysis of case studies, resulting in the development of an information database. This stage supports applied research by connecting empirical observations to initial propositions and helps to define appropriate investigative tools (Yin 1981; Cross 2001). WP3 marks the applicative phase to test and validate the research approach through a tangible output. WP4 addresses the analytical assessment of assumptions and findings, aiming to synthesize the explored areas and identify potential applications and future implications.

We Should Stop Building New Stuff

The building industry is responsible for around 40 percent of global carbon dioxide emissions,[10] a fact that raises serious questions about how we define progress today (Germanà 2025). Modern society is driven by constant innovation and growth, with the belief that the tools we use at any given moment will soon be outdated[11] (Mollick 2024). Indeed, technology is always expected to improve, and this reality is tied to a capitalist model that depends on steady profits, often achieved through faster and more advanced technologies. But these gains come at the cost of natural resources, which are continually used up in the name of development (Haraway 2015). The Earth becomes an object of consumption, or as Charlotte Malterre-Barthes would argue, a "commodity" for the sake of expanding the built environment (Malterre-Barthes 2024), where the mass of human-made materials has already surpassed that of all natural biomass (Elhacham et al. 2020).

This manuscript focuses on restoration, not just as a technical act, but as a cultural and disciplinary shift that embraces a "labor of repair." While the narrative looks at specific practices within the conservation of cultural heritage, the broader aim is to reconsider how technology might help us care for the built world we already have. Instead of focusing on building more, we should prioritize repairing and rethinking what already exists. There are many underused resources, including existing buildings, that are left unattended for too long and are an untapped source of potential. As of 2020, the global built floor area reached an estimated 13 billion square meters, or about 139 billion square feet (Zhang et al. 2024). Yet this built space is not equally distributed, and much of it is not even fully used. As an example, largely due to the cultural affirmation of hybrid work, nearly 93 million square meters, or a billion square feet, of office space is now vacant in the United States. When considering building performance, according to the European Commission's report *A Renovation Wave for Europe*, over 220 million buildings in the European Union, roughly 85 percent of the total, were built before 2001. Most of these exhibit poor energy efficiency, and 85 to 95 percent of them are expected to still be in use by 2050. Buildings are responsible for 36 percent of the EU's greenhouse gas emissions from energy, but only 11 percent of the existing building stock is renovated each year. This is a lost opportunity to remediate existing conditions into more efficient versions of themselves. Deep renovations that cut energy use by at least 60 percent occur in just 0.2 percent of buildings. Meanwhile, the construction industry remains a major driver of raw material extraction and global warming (O'Donnell and Pranger 2020), and construction waste accounts for about one-third of all solid waste, further exacerbating a growing problem. At this rate, the crisis we face will take centuries to resolve. This is not just a question of scale but of the fundamentally extractive nature of the discipline of architecture. So perhaps the real question is not how to come up with new workflows or rethink architectural restoration theory, but how to shift the entire mode of practice toward a necessary focus on recovery, reuse, and care to ensure that "buildings persist in time" (Mostafavi and Leatherbarrow 1993).

To move forward responsibly, we should challenge long-standing norms: re-define standards, and look more deeply at how supply chains function, how labor is organized, and how real estate shapes what gets built or maintained. This kind of shift also demands new strategies for maintenance and stewardship, placing care for the collective built environment within a framework that claims to be humanist but must now also take on environmental, material, and social responsibility. Addressing today's challenges requires confronting questionable assumptions that building more is an effective solution and considering, when the value of "doing nothing" (Odell 2019) would be successful. Just as important, we need to break down the boundaries that separate these fields of knowledge. The future lies in the fabric of a new kind of material literacy; one that embraces repair, uses digital and mechanical technologies to support rather than replace labor, minimizes advantages of demolition over renewal,[12] and reinforces care and humanity into the discipline of architecture.

Notes

1 The term "palimpsest" describes a constructed environment that reveals visible layers of its historical development, where elements from past eras remain embedded within newer additions. Much like a manuscript that has been partially erased and overwritten, yet still bears faint traces of the original text, such a structure carries the imprints of its evolving past beneath the surface of the present.

2 "Technocratic" describes an approach that privileges technical expertise and scientific principles in the design and construction process, placing efficiency, functionality, and technological application above aesthetic or subjective considerations. In this study, technology should not be regarded as the solution in itself but as a partner that supports and enables humanistic design decisions.

3 The concept of "Industry 4.0" positions itself as the next phase in a series of transformative technological shifts—following the use of water and steam power (1.0), electric power (2.0), and computing and automation (3.0). This perspective is explored by Christian Fuchs in "tripleC: Communication, Capitalism & Critique."

4 UNESCO stands for United Nations Educational Scientific and Cultural Organization.

5 UNESCO List of World Heritage in Danger: https://whc.unesco.org/en/danger-list/.

6 Conservation refers to practices aimed at safeguarding the material integrity and historical authenticity of a building. In this sense, it differs from renovation, reuse, and reconstruction, which prioritize functional adaptation or visual renewal over the continuity of original materials and fabric. Functional interventions, such as the addition of elevators, ramps, or staircases to enable reuse, typically fall outside the scope of conservation and do not necessarily require formal recognizability.

7 For completeness, it is worth noting that the distinction between restoration and conservation has led to further theoretical interpretations. Some perspectives see them as distinct methodologies but still assert that architects should oversee all architectural interventions, even those focused solely on conservation.

8 From a seminar on architectural drawing in archaeology delivered by Francesca Eugenia Damiano, Department of Cultural Heritage, University of Macerata, 2020.

9 George Devol's patent, "Programmed Article Transfer" (U.S. Patent No. 2,988,237), represents the first reprogrammable robotic arm.

10 Analysis by Architecture2030 using data sources from IEA & Statista.

11 In *Co-Intelligence: Living and Working with AI*, Ethan Mollick offers a set of principles for engaging with artificial intelligence. Amid rapid technological acceleration driven by trained algorithms, he suggests we should assume "this is the worst AI you will ever use," referring to the continuous, exponential improvement of these systems.

12 Slonimskiy, Leonid, et al. "Forced Reuse: Rethinking Building Conventions through Crises." *e-flux Architecture*, June 2025, www.e-flux.com/architecture/framing-renovation/590375/forced-reuse-rethinking-building-conventions-through-crises/.

References

Abelson, Harold, Gerald Jay Sussman, and Julie Sussman. 1996. *Structure and Interpretation of Computer Programs*. 2nd ed. MIT Press.

Alleyne, Brian. 2019. *Geek and Hacker Stories: Code, Culture and Storytelling from the Technosphere*. Palgrave Macmillan UK. https://doi.org/10.1057/978-1-349-95819-1.

Booth, Wayne C., William C. Booth, Gregory G. Colomb, Gregory G. Colomb, Joseph M. Williams, and Joseph M. Williams. 2003. *The Craft of Research*. University of Chicago Press.

Brandi, Cesare. 1963. *Teoria del Restauro*. Edited by Licia Vlad Borrelli and Joselita Raspi Serra. Ed. di Storia e Letteratura.

Brynjolfsson, Erik, and Andrew McAfee. 2014. *The Second Machine Age: Work, Progress, and Prosperity in a Time of Brilliant Technologies*. WW Norton & Company.

Carbonara, Giovanni. 2013. *Il Restauro non è Conservazione*… Lactiones Magistrales. Sapienza università di Roma, Facoltà di Architettura.

Carpo, Mario, ed. 2013. *The Digital Turn in Architecture 1992–2012*. 1st ed. Wiley. https://doi.org/10.1002/9781118795811.

Carpo, Mario. 2017. *The Second Digital Turn: Design Beyond Intelligence*. With Cynthia Davidson. Writing Architecture Series. The MIT Press.

Creswell, John W., and J. David Creswell. 2017. *Research Design: Qualitative, Quantitative, and Mixed Methods Approaches*. Sage Publications.

Cross, Nigel. 1984. *Developments in Design Methodology*. Wiley Chichester.

Cross, Nigel. 2001. "Designerly Ways of Knowing: Design Discipline versus Design Science." *Design Issues* 17 (3): 49–55.

Dalla Negra, Riccardo. 2017. "Architettura e Preesistenza: Quale Centralità?" In *Architettura e Preesistenze. Premio Internazionale Domus Restauro e Conservazione Fassa Bortolo*. Skira.

Daubmann, Karl. 2004. "Teaching Digital Fabrication through Design." *Fabrication: Examining the Digital Practice of Architecture*, November 8, 244–55.
Davis, Stanley M. 1997. *Future Perfect*. Basic Books.
Elhacham, Emily, Liad Ben-Uri, Jonathan Grozovski, Yinon M. Bar-On, and Ron Milo. 2020. "Global Human-Made Mass Exceeds All Living Biomass." *Nature* 588 (7838): 442–44.
European Commission. 2024. *The Future of Europe's Past: Why Member States Must Do More to Advance Digitisation of Our Cultural Heritage: Implementation of the 2021 Commission Recommendation on a Common European Data Space for Cultural Heritage: Progress Report 2021 2023*. Publications Office. https://data.europa.eu/doi/10.2759/5921820.
Figliola, Angelo. 2017. "Post-Industrial Robotics: Exploring Informed Architectures in the Post-Digital Era." *TECHNE – Journal of Technology for Architecture and Environment*, August 4, 256–66. https://doi.org/10.13128/TECHNE-19751.
Foster, Richard N. 1986. *Innovation: The Attacker's Advantage*. Summit Books.
Germanà, M. L. 2004. *Significati dell'Affidabilità Negli Interventi Conservativi*.
Germanà, Maria Luisa. 2025. "Beyond or within the Limits? A Misleading Dilemma for Current Technological Processes." *TECHNE-Journal of Technology for Architecture and Environment*, 3, 25–28.
Gershenfeld, Neil. 2012. *Foreign Affairs: How to Make Almost Anything: The Digital Fabrication Revolution*. Vol. 91.
Groak, Steven. 2002. *The Idea of Building: Thought and Action in the Design and Production of Buildings*. Taylor & Francis.
Hackett, Sophie, Andrea Kunard, Urs Stahel, et al., eds. 2018. *Anthropocene: Burtynsky, Baichwal, de Pencier*. Art Gallery of Ontario; Goose Lane Editions.
Haraway, Donna. 2015. *Simians, Cyborgs, and Women: The Reinvention of Nature*. Routledge.
Hobsbawm, Eric J., and Terence O. Ranger, eds. 2012. *The Invention of Tradition*. Canto Classics. Cambridge University Press.
Kieran, Stephen, and James Timberlake. 2004. *Refabricating Architecture: How Manufacturing Methodologies Are Poised to Transform Building Construction*. McGraw-Hill.
Kolarevic, Branko. 2004. *Architecture in the Digital Age: Design and Manufacturing*. Taylor & Francis.
Leach, Neil. 2002. *Designing for a Digital World*. John Wiley & Sons.
Malterre-Barthes, Charlotte. 2024. "Teaching Not To: 'Stop Building' as a Brief for the Architecture School of the Present." *Journal of Architectural Education* 78 (1): 158–75. https://doi.org/10.1080/10464883.2024.2303932.
Mollick, Ethan. 2024. *Co-Intelligence: Living and Working with AI*. Portfolio/Penguin.
Mostafavi, Mohsen, and David Leatherbarrow. 1993. *On Weathering: The Life of Buildings in Time*. MIT Press.
Odell, Jenny. 2019. *How to Do Nothing: Resisting the Attention Economy*. Melville House.
O'Donnell, Caroline, and Dillon Pranger. 2020. *The Architecture of Waste: Design for a Circular Economy*. Routledge.

Otero-Pailos, Jorge, Erik Langdalen, and Thordis Arrhenius, eds. 2016. *Experimental Preservation*. Lars Müller Publishers.
Picon, Antoine. 2014. "Robots and Architecture: Experiments, Fiction, Epistemology." *Architectural Design* 84 (3): 54–59. https://doi.org/10.1002/ad.1754.
Schwab, Klaus. 2017. *The Fourth Industrial Revolution*. Currency.
Sharma, Pranjal. 2023. *The Next New: Navigating the Fifth Industrial Revolution*. Harper Business, an imprint of HarperCollins Publishers.
Snyder, James C., ed. 1984. *Architectural Research*. Environmental Design Series 6. Van Nostrand Reinhold.
Spadolini, Pier Luigi. 1981. *Progettare nel Processo Edilizio: La Realtà Come Scenario per l'Edilizia Residenziale*. Edited by Mario Zaffagnini. Luigi Parma.
Tafuri, Manfredo. 1991. "Storia, Conservazione e Restauro: Una Intervista a Cura Di Chiara Baglione e Bruno Pedretti." *Casabella: Rivista Mensile Di Architettura e Tecnica*, no. 580: 23–26.
Torsello, Paolo, ed. 2005. *Che Cos'è il Restauro? Nove Studiosi a Confronto*. 1st ed. Elementi. Marsilio.
UNESCO. 2004. *Proceedings of the International Conference on the Safeguarding of Tangible and Intangible Cultural Heritage: Towards an Integrated Approach*. United Nations Educational, Scientific and Cultural Organisation.
Yin, Robert K. 1981. "The Case Study as a Serious Research Strategy." *Knowledge* 3 (1): 97–114.
Yin, Robert K. 2017. *Case Study Research and Applications: Design and Methods*. Sage Publications.
Zhang, Shufan, Minda Ma, Nan Zhou, et al. 2024. "Estimation of Global Building Stocks by 2070: Unlocking Renovation Potential." *Nexus* 1 (3).

Chapter 2

Digitalization of Building Processes

DOI: 10.4324/9781003590873-2

Fourth Industrial Revolution and Role of Digital Tools in Construction

The term "Industry 4.0" was introduced for the first time in 2011 at an event called the *Hannover Fair* to describe an ongoing global change dictated by connectivity and automation of production tools.[1] Connectivity and automation are the foundation of smart factories, manufacturing environments where the physical and digital realms are seamlessly integrated. In 2014, Erik Brynjolfsson and Andrew McAfee introduced the concept of the Second Machine Age, describing a new era in which machines gained cognitive autonomy through the rise of cyber-physical systems. This period marks the point when digital technologies had matured enough to drive widespread social and economic transformation (Brynjolfsson and McAfee 2014). In 2015, Klaus Schwab, founder of the World Economic Forum, introduced the concept of the Fourth Industrial Revolution, describing it as a "fusion of technologies blurring the lines between the physical, digital, and biological spheres." This definition captures the accelerating pace of change in contemporary culture, particularly its widespread effects on economics, education, and work, with the biggest impact on manufacturing (Fig. 2.1). In his book, *The Fourth Industrial Revolution*, Schwab outlines the developments that led to this global transformation, emphasizing that its speed and scale surpass those of previous industrial shifts (Schwab 2017).

Around 10,000 years ago, the domestication of animals and the shift from foraging to farming fundamentally changed human life leading to larger, more permanent settlements and the emergence of organized food production. This transformation supported local population growth, eventually leading to the birth of cities and early forms of urbanization. Centuries later, the First Industrial Revolution emerged within these urban centers, as muscle power was replaced by mechanical power following the invention of the steam engine in the 1760s. Over the following 120 years, industrialization spread across Europe, and by the early 20th century, the advent of electricity laid the groundwork for the Second Industrial Revolution with the development of assembly lines, enabling mass production

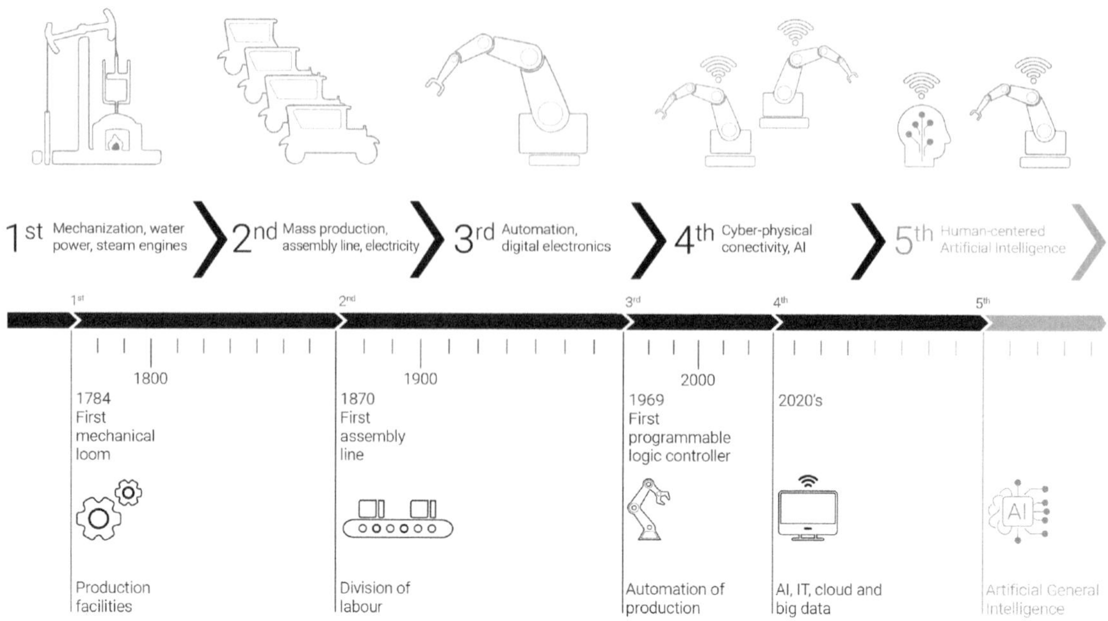

Figure 2.1 Timeline of the Industrial Revolutions: the First (1760–1840), Second (late 19th to early 20th century), Third (1960s), and Fourth (early 21st century to present), with a projected emergence of a Fifth Industrial Revolution. Diagram adapted from Christoph Roser at *AllAboutLean.com*, licensed under CC BY-SA 4.0.

and the rise of economies of scale. Looking ahead, the emergence of computers in the 1960s, the rise of personal computing in the 1970s and 1980s, and the introduction of the internet in the 1990s established a digital ecosystem that propelled the onset of the Third Industrial Revolution, which has yet to be fully exploited because as recently as 2024, there are still 2.5 billion people that lack internet access and connectivity.[2]

In the 2010s, the convergence of mobile internet, exponential computing power, the availability of vast data sets, the democratization of automated production tools, advanced sensor technologies, and machine learning catalyzed the Fourth Industrial Revolution. This phase has accelerated the dominance of digital business models, which operate through online platforms and, as Jeremy Rifkin noted, have "reduced the marginal cost derived from transactions for the production of products, goods, or services" (Rifkin 2014). Internet transformations inform means of production. Tracking down these developments, the world went from the non-standard seriality of the static, read-only Web 1.0, to the collective intelligence of the Web 2.0 of the late 1990s. Next was the semantic Web 3.0 of the early 2000s, which progressed into our current connective abilities shown in Web 4.0, where we are able to connect all devices across real and virtual worlds in real time. Connectivity 4.0 allows machines to communicate with one another through open networks and semantic descriptions. Looking ahead, Web 5.0, referred to as the emotional or sensory web, remains speculative, imagining a future where digital systems can respond empathetically by recognizing and interpreting human emotions.

In this framework, designers have the opportunity to access new physical and digital tools through which to expand the creative possibilities and enable the customization of products as an expression of the post-industrial and post-Fordist era of ready-made products.[3] As Mario Carpo states in *Beyond Digital*, for over a 100 years, we attempted to transform building into a factory, hoping to mechanize construction using the tools of industrial mass production. That approach never succeeded. Understandably, we are now pursuing the opposite: using intelligent robotics to develop a neo-artisanal mode of making (Carpo 2023b), where voxels[4] (three-dimensional units) interact with pixels (two-dimensional data).

Consequences in Architecture

The Second Digital Turn

The Fourth Industrial Revolution is more closely associated with transformations in manufacturing, whereas the Second Digital Turn carries distinct nuances specific to the field of architectural design. The path to this point began in the early 1990s with the introduction of digital tools in architecture, many of which originated in the aerospace and aeronautics sectors. The introduction of digitization in the field of architecture, termed Digital Turn (Carpo 2013), favored the creation of new formal languages, "the new digital style of smooth and curving, 'spliny' lines and surfaces" (Carpo 2017) that moved beyond the fragmented aesthetics of deconstructivism. Form generation could now take place within virtual environments through two- and three-dimensional simulations of both form and material behavior leading to more flexible design approaches. This was further supported, from the 2000s onward, by the development of automated tools—such as robotic arms, extruders, and 3D printers—capable of translating digital models into physical objects, circumventing the limitations of traditional, standardized production methods. This evolving design paradigm, based on the integration of computation and materialization, was theorized by Mario Carpo as the Second Digital Turn (Carpo 2017).

The logic of digital tools encourages project generation, project description, and materialization to be intertwined. Digital fabrication and robotics allow for design and construction to occur simultaneously, dissolving the traditional separation between conception and execution. As a result, the role of the architect, rooted in the Renaissance ideal of elevating ideas over hand-on practice, is transformed through CAD-CAM systems that bring together thinking and making. Both processes are now carried out using the same tool: the computer. Computers can formulate a vast number of hypotheses and evaluate them rapidly until the most suitable or advantageous solution is identified, enabling the exponential generation of highly complex structural elements that would be difficult to conceive or test manually (Bonwetsch et al. 2010). From this perspective, "in a sense, the use of a computer is more medieval than modern,"[5] as it simultaneously manages dimensional, geometric, numerical, and structural information. In this way, contemporary digital practice recalls the working methods of the pre-renaissance ages, when the master builder made critical decisions directly on the construction site, integrating design and execution in real time (Fig. 2.2).

In *The Second Digital Turn,* Mario Carpo recalls how Leon Battista Alberti moved forward the idea that architectural production should be expressed with drawings, or written notations, and then transformed by builders as identical copies in reality. The principle of architectural authorship is therefore inherent in its representation and becomes tangible once

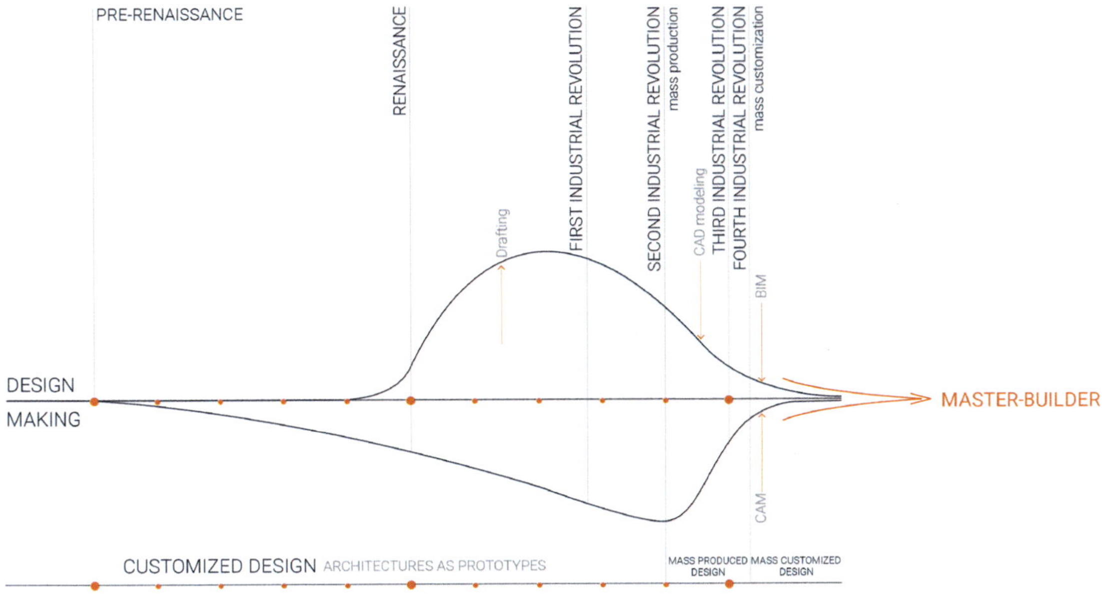

Figure 2.2 Contraction and deviation in design and making through industrial revolutions. Before the industrial revolution, every building was effectively a prototype, a unique materialization of an idea. Design solutions were non-standard and defined case by case, embedding the intelligence of accumulated techniques carried forward by craftspeople.

the geometric annotation is transferred to the physical world through construction work. During the 2017 Fabricate conference, Mario Carpo revisited this historical shift, noting that we may now be returning to a pre-modern model of making. He described this as a form of "digital craft," explaining that "we are much closer to the way we made physical things 500 or 600 years ago," before the Renaissance separation of design and construction. In that earlier context, master builders conceived and executed their work simultaneously, making design decisions directly on site. "Paradoxically," Carpo concluded, "we are returning to this" (Sheil et al. 2017).

In a more recent essay, Mario Carpo summarized what he calls a "short but believable history of the digital turn in architecture," highlighting the growing role of computation in design, including the resurgence of artificial intelligence (Carpo 2023a). AI, a technology with roots in the late 1950s, is now re-emerging with increased relevance in tasks focused on problem-solving and decision-making. However, he cautions that data-driven AI remains limited to quantifiable phenomena, while architectural design, rich in cultural, spatial, and qualitative dimensions, resists easy translation into numerical terms. This perspective tempers the notion of a third digital turn led by AI (Carpo 2023b), instead reinforcing the idea that we are still navigating a post-notational paradigm of human intelligence that may no longer be entirely human, yet remains grounded in forms of authorship that transcend pure computation.

Fabrication-Aware Design

The Fourth Industrial Revolution defines a broad technological and cultural ecosystem, while the Second Digital Turn represents its specific manifestation within the field of architecture. In this framework, fabrication-aware design emerges as a distinct design approach that integrates material production into the digital design process. Fabrication-aware design is the ability to inform digital models so that they can "generate buildable formal solutions" (Pottmann 2013). This approach embeds fabrication considerations from the earliest, ideational stages of a project, helping to define the limits of formal exploration and ensuring the constructability of the resulting components.

The evolving relationship between design and building, bridging conception and fabrication, was already addressed by Branko Kolarevic in 2004. In *Architecture in the Digital Age,* he notes, "in the future, being an architect will also mean being a builder, not literally, of course, but by digitally generating information to manufacture and construct buildings in ways that render present inefficient hierarchies of intermediation unnecessary" (Kolarevic 2004). Computational design supports this shift through a holistic framework for structuring design parameters and data hierarchies. Geometry, physical simulations, and material behavior are connected into a unified digital environment. Under these conditions, the dimensional limitations of fabrication tools become design constraints that inform the generation of form. Design intelligence transferred into the model expands the spectrum of possibilities in digital design, grounded in real-world production logic.

The fabrication-aware design process makes it possible to evaluate whether it is more efficient to produce customized off-site components, pursue on-site fabrication, or implement a hybrid prefabrication strategy. This mode of practice also opens future possibilities for on-site robotic construction, which could significantly shorten the production chain and reduce transportation costs. Furthermore, this design methodology supports the integration of two complementary design streams: downstream or top-down approaches driven by form and program, and upstream or bottom-up approaches focused on components and fabrication. When downstream design decisions are informed by production methods, the process enriches early creative phases with embedded technical knowledge and enables a more integrated, cognition-driven workflow.

Digitization, Digitalization, and Digital Transformation

Digital involvement in a process, product, or project can occur through various modalities and at different scales. Technological adoption (Fig. 2.3), intuitively, serves as an indicator of how disruptive

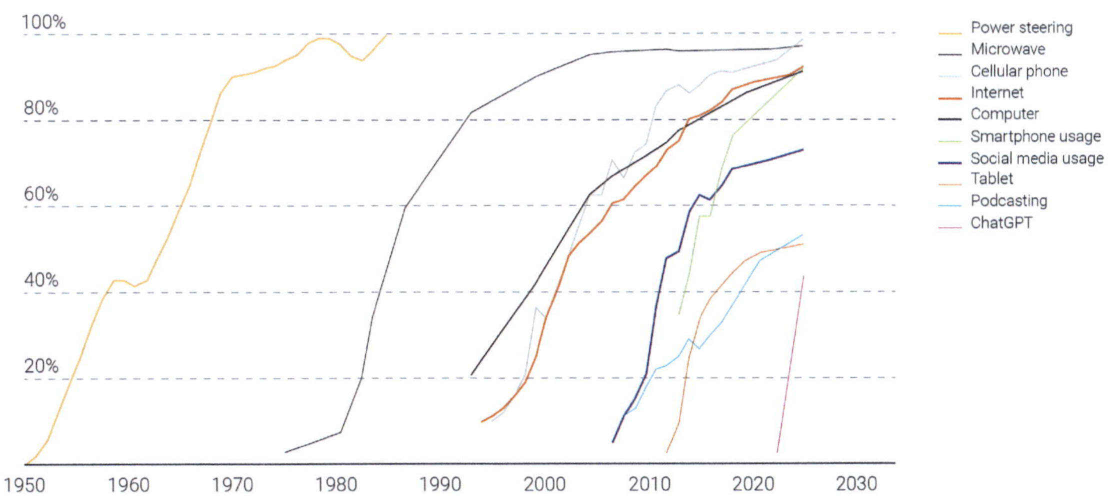

Figure 2.3 Curves illustrating the speed of technology adoption in the United States. Reworked from source: *Visual Capitalist.*

“ **Digitization** is the act of converting information from its original analog form into a digital version.	CAD
“ **Digitalization** describes how organizations apply digital technology to evolve their models and open up new avenues for growth.	BIM
“ **Digital transformation** is a strategy-driven shift focused on user needs and enabled by digital technologies.	WORKFLOW

Figure 2.4 Digitization, Digitalization and Digital Transformation as modes of practice in architecture.

a technology may be as a driver of change impacting multiple aspects of society, from the organization of work to modes of production, business models, and education. One effective way to understand complex ecosystems is to break down broader challenges into smaller components. Understanding the nuanced differences between digitization, digitalization, and digital transformation allows us to better grasp the mechanics behind technological change (Fig. 2.4).

According to *Gartner's IT Glossary*, digitization is defined as “the process of changing from analog to digital form, also known as digital enablement.” This process consists of transferring analog information or tasks into a digital format, allowing for their representation and manipulation in the virtual realm. In the context of architecture, digitization refers to the introduction of CAD (computer-aided design) tools, digital interfaces that replicate traditional drafting processes and make it possible to create shapes and drawings on screen rather than by hand.

Digitalization, on the other hand, is described as “the use of digital technologies to change business models and provide new revenue and value-producing opportunities.” In architecture and construction, this is exemplified by building information modeling (BIM), an operational tool that not only facilitates design but also integrates the management of project timelines, budgets, and construction phases. Through digitalization, processes once handled separately are now unified within a single, data-driven environment.

Finally, digital transformation refers to a strategic shift driven by user needs that integrates digital technologies across all aspects of an organization. This advanced digital ecosystem is powered by digital tools, data analytics, artificial intelligence, and machine learning. As a result, modes of operation evolve in tandem with shifts in technology, culture, and workforce dynamics.

In architecture, digital transformation redefines conventional workflows, allowing design, planning, and construction activities to unfold concurrently, instead of following a linear sequence. Digital transformation informs decision-making at every level, placing greater emphasis on analytics and the use of data to guide strategy and execution. As workflows become increasingly reliant on information technologies capable of storing, processing, and leveraging data, new skill sets are required across sectors.[6] This shift will have a significant impact on education, particularly in higher education, which is expected to lead this transformation.[7] There is a growing need to cultivate both technical competencies and a cultural mindset that can support evolving practices,[8] especially in preparation for roles that have yet to emerge within an increasingly digital and fast-changing economy.

Together, digitization, digitalization, and digital transformation mark a progression from digital representation to operational optimization to a redefinition of professional practice. Understanding this continuum is a design prompt for architects and educators to navigate the future of the built environment.

Digital Transfer and New Project Culture

Bringing digital tools into construction means learning from industries that adopted technology earlier. Architecture, described by Greg Lynn as a big assembly of components that need to be organized,[9] has long looked to aerospace and naval design,[10] sectors that integrated digital processes well before architecture began to do so. By the 1990s, cars, planes, and ships were already being fully conceived, developed, tested, and refined within digital environments, as their production relied on computer-controlled technologies. The concept known as "file-to-factory" originated in aeronautics, where CAD models feed directly into CNC machinery, to help control consistency and identify production efficiencies. Transportation design contends with more than just gravity. Aerodynamic and hydrostatic forces demand smooth, continuous surfaces that cannot be described with simple geometry. To describe the continuous, complex surfaces required in transportation design, calculus was adopted as the underlying mathematical principle to describe geometry (Lynn 1999). In the 1960s, Pierre Bézier and Paul de Casteljau, while at Renault and Citroën, developed methods to work with calculus and make design repeatable, moving beyond the artisanal way of trial and error (Brennan 2016). Their work introduced the term "spline" into modern design vocabulary, a mathematical function to describe a smooth curve, originally named after a flexible wooden strip used in shipbuilding to shape hulls. This foundational concept later evolved into more advanced mathematical models, leading to the development of non-uniform rational B-splines (NURBS), which underpin most contemporary 3D modeling software. Before this digital shift, workflows depended on physical scale models, manually measured and translated into CAD. Today, it's more common to produce a CNC-milled clay model to quickly test and refine digital forms through affordable iterations.

In the aircraft industry, full-scale drawings were once used to design planes before digitization. Willow Run, an automotive Ford plant in Michigan,[11] assisted the US government during World War II by building airplanes. Inside the plant, curvilinear elements, now modeled digitally using NURBS, were produced by hand, crafted by skilled workers without the aid of digital modeling. A visual record of this process is captured in the iconic photo of the drafting room at the bomber factory by Hedrich Blessing Studio, showing a draftsperson in the foreground bending a wooden strip with bronze weights to draw an analog spline.[12] This reliance on labor-intensive methods began to shift in the 1980s, when the French military aviation company Dassault Aviation invested in digital systems to streamline production. The company developed computer-aided three-dimensional interactive application (CATIA), a CAD program that transitioned design work from two-dimensional drafting to fully integrated three-dimensional modeling. CATIA's capacity to link design with fabrication transformed industrial workflows. One of its best-known test cases was the creation of the Boeing 777, the first aircraft to be entirely designed and tested within a digital environment from concept to production (Kolarevic 2004).

While construction practices have seen only minor changes over the past two centuries, a few forward-looking examples anticipated a shift. Buckminster Fuller applied aerospace methodologies in the design of the Dymaxion car and house,[13] combining engineering logic with architectural vision. The transfer of parametric input-output design workflow into architecture became more defined with Frank Gehry's use of CATIA. His *1992 Fish Sculpture* in Barcelona was the first architectural-scale project realized without traditional two-dimensional drawings, relying instead on a fully digital, paperless process. Gehry used CATIA to manage the translation of complex forms from model to construction and later applied the same approach to the Guggenheim Museum in Bilbao. He has since remained a leading advocate for digital fabrication in architecture and for the file-to-factory workflow, which has reshaped professional roles by bringing architects into more direct contact with the construction process.

Building on these examples, a new project culture emerged, where the materialization of complex virtual spaces and components through digital fabrication expanded the range of design possibilities and overcame creative limitations previously imposed by traditional representation and manufacturing tools. When fabrication strategies align with formal generation techniques, architectural ideas can be shaped in direct relation to material properties from the earliest stages of the design process. This reduces both the physical and procedural distance between the designer and the act of making, supporting a renewed understanding of architecture as a material practice (Figliola 2017).

Connections Between Technology and Architectural Heritage

Architectural conservation has long relied on specialized labor, which has slowed its adoption of technological advances, particularly those related to material applications, beyond data capture and surveying. However, conservation is not limited to historic architectural heritage but also addresses more recent buildings that incorporate industrial materials and processes. As a result, the discipline has had to pay attention to industrial developments in order to understand how these buildings, products of modern production systems, will require restoration and care in the centuries to come. For instance, in the early 20th century, the mass production of consumer goods extended to the construction industry, introducing standardized and cost-effective building components, with new materials being created and the performance of existing materials improved.

In this context, modernists laid the foundations for a new idea of architecture, one that was rational, functional, and uninterested in ornamentation. There was a growing awareness in giving aesthetic dignity to prefabricated materials that were the result of mass production, celebrating their form and textural qualities, like in Joseph Paxton's emphasis on iron and glass in the Crystal Palace, Le Corbusier's use of concrete, Mies Van Der Rohe's incorporation of steel, and Louis Kahn favoring the use of brick, the archetypal prefabricated material. This period marked a clear break from traditional construction and aesthetic values. Restoration practice was forced to respond, broadening its strategies and engaging with a new repertoire of formal approaches to address interventions in culturally, historically, or architecturally significant buildings, a challenge that remains open and continually evolving.

The practice of restoration has already benefited from digital tools for surveying, such as laser scanners, cameras, and sensors, and from diagnostic equipment including endoscopes, ultrasound, and infrared devices. These technologies support data capture, photogrammetry, point cloud management, 3D modeling, and high-fidelity digital reproduction. On-site, these tools can now communicate with one another, updating collected data in real time to reflect the evolving condition of a building.

It has also become common practice to use robotic devices and drones to perform technical and cognitive tasks during architectural analysis, particularly in hazardous or inaccessible areas, such as those affected by structural collapse.

Another development has been the adaptation of BIM to existing buildings. BIM introduced optimized digital models that function as transparent documents containing all data related to a building's entire life cycle. In this type of process, the informative character of the model becomes predominant over the geometry, with the model itself serving as a representation, simulation, or software output, from which architectural elements and technical documentation can be efficiently extracted (Aish 2013).

From this foundation emerged Heritage BIM (HBIM), a methodology specifically developed for the documentation, conservation, and management of historic structures, enabling control over processes from investigation to restoration (Quattrini et al. 2015). The term HBIM was first introduced in 2009 by Maurice Murphy and colleagues at the Dublin Institute of Technology (Murphy et al. 2009). Unlike standard BIM, which is typically used for new buildings with standardized components, HBIM is designed to handle the irregular geometries and layered historical complexity of heritage architecture. This software environment combines high-resolution 3D modeling with metadata related to materials, construction phases, historical context, and conservation requirements. HBIM prioritizes the semantic organization of information and focuses on identifying the types of data necessary to guide effective conservation strategies, beginning with accurate representation (Germanà 2019).

Recent experimentations have extended HBIM into broader domains of spatial computing,[14] integrating extended reality, the Internet of Things, artificial intelligence, and machine learning. These technologies aim to create systems that respond to the physical environment in real time while merging multidisciplinary expertise (Fig. 2.5), tacit knowledge (Otero 2024), and layered information. One example is the Heritage++ project developed at ETH Zurich, which explored the creation of collaborative

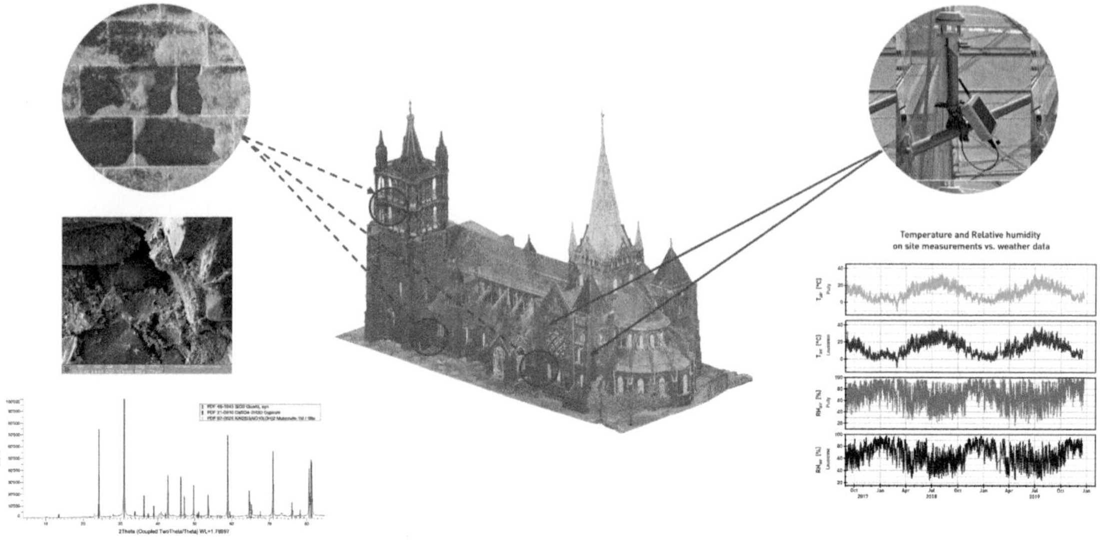

Figure 2.5 Schematic illustration of the spatial and multiscale challenges involved in assessing degradation mechanisms affecting a monument in order to define effective conservation measures. The Lausanne Cathedral is used as a case study, showing how three-dimensional relations between material types (petrology, microscopy, mineralogy) and microclimatic conditions (wind direction, temperature, humidity, and driven rain) can be visualized. Courtesy of Yamini Patankar.

platforms and interactive environments where experts can intuitively access, analyze, and share complex datasets (Patankar et al. 2024). Through a user-friendly interface, conservation professionals can retrieve the most relevant information to support decision-making and guide conservation strategies, including the use of AR to assess structural building performance (Fig. 2.6).

However, the technologies associated with the Fourth Industrial Revolution have yet to fully demonstrate their potential in on-site applications or in work with experimental materials. This is due not only to the challenges of implementation but also to the difficulty of scaling such approaches from both technical and theoretical standpoints. Attempts are emerging, however, in the context of digital restoration, an area that applies advanced digital technologies to historical reconstruction and preservation. This is evident in the work of practices such as Factum Arte in Spain and the Institute of Digital Archaeology in the UK, as well as in

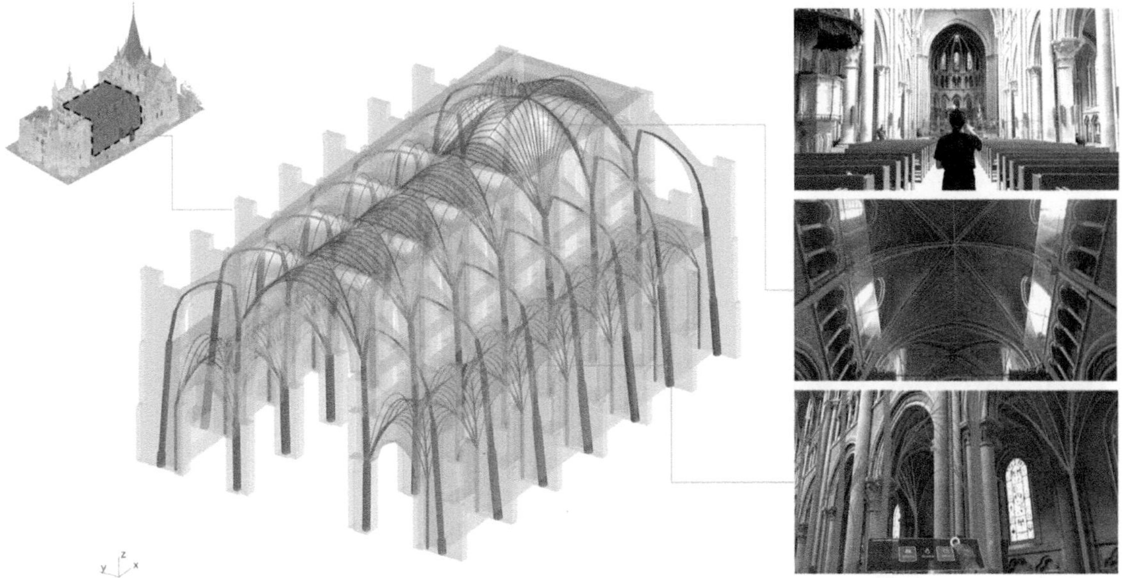

Figure 2.6 Global structural analysis of the Lausanne Cathedral illustrating the flow of compressive forces within its main elements, alongside an AR platform that enables users to visualize and interact with the results on-site. Diagram by Ricardo Maia Avelino.

academic research that focuses on creating faithful replicas of historical elements. The typical methodology of digital restoration follows a linear multi-step process: reproduction of the element, off-site prototyping (often via CNC milling or 3D printing), post-processing, and final on-site fitting (Rudin et al. 2023).

Future developments are expected to shift these methods toward on-site automation and advanced experimentation, evolving beyond the replication of individual components toward more holistic forms of intervention. This approach would integrate tools, workflows, and cultural shifts to define new design languages, likely shaped by parametric customization and novel materials emerging from ongoing research. Operationally, this future-oriented design practice would be optimized through AI, sensors, and predictive models trained with machine learning, linking digital data streams to on-site tasks. These context-aware computing systems would respond to the existing built environment, enabling data-informed decision-making throughout the restoration process. Achieving this shift depends on a deep understanding of how to establish reliable operational methodologies for any conservation or restoration need, addressing both tangible and intangible dimensions. On the one hand, this involves making appropriate material choices; on the other, it requires structuring the intervention phases systematically, minimizing project uncertainties (Germanà 2019), and clearly defining the necessary skills, tools, and operating costs to carry out preservation, conservation, and recovery operations.

Democratization of Robot Scripting

Industrial robotic arms are now widely used in research centers and academic institutions. This is largely due to the progressive democratization of robotics, both in terms of hardware and software. Manufacturers are increasingly producing robots specifically for educational purposes, and it has also become possible to purchase used industrial robots from decommissioned assembly lines. At the same time, programming interfaces have become more accessible, simplifying communication between digital models and manufacturing tools.

CAD-CAM processes have become simpler and faster thanks to a direct relationship with digital fabrication tools that now share the same physical and digital space as the designer. Smart robots, a category of machines equipped with sensors, serve not only as extensions of the workspace but also as extensions of the designer's cognitive environment. In this working context, production is shaped by the designer's competence, and the manufacturing phases can be managed within the same digital environment where the project's form is conceived. This integration allows for new design possibilities and the development of customized workflows that retain the efficiency typically associated with serial production.

Today, it is possible to program industrial robots using the same tools employed in architectural design. Beyond proprietary software, robots such as those made by Kuka and ABB can be controlled within the Rhinoceros and Grasshopper algorithms-aided software environment (Tedeschi 2014) through dedicated plugins like Robot Components or Kuka|PRC.

These plugins allow users to define design parameters alongside coordinated robotic motion. For instance, Kuka|PRC was first released in 2011 by Johannes Braumann, co-founder with Sigrid Brell-Cokcan of the Association for Robots in Architecture, a community dedicated to advancing research in the field. Using this parametric design logic, robotic programming operates as a bridge between digital concepts and physical fabrication. Once the informative model, or Grasshopper definition, is established, the robot becomes an active agent linking design intent to material execution.

When programming robotics, what is actually being designed is the robot's kinematics, the motion of the robot through its typical six or seven axes. These are defined based on the type of actuator installed on the robot, the material being manipulated, and the overall design objective. Movements are programmed by setting a sequence of target points in the three-dimensional space, each with a specific position and orientation. The arrangement of these points is known as toolpathing (Fig. 2.7).

Despite the precision of robotics and the coordination of sequencing, the process still demands a certain level of craft and orchestration with the

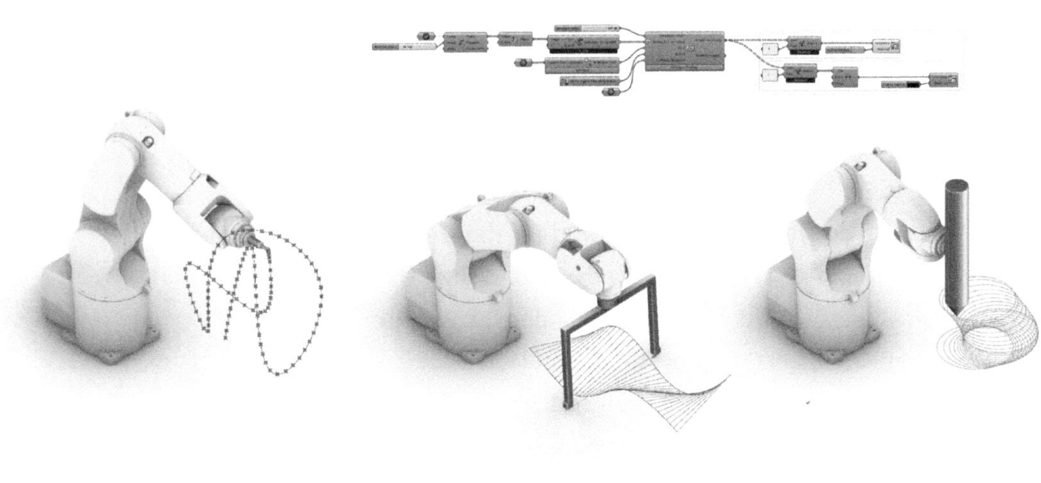

Figure 2.7 Algorithm-aided design enables robots to be programmed for various tasks using different end effectors. A 3D preview visualization supports early detection and correction of kinematic errors before executing operations in the physical environment.

tools. Traditional material practices are translated into the control of digital variables, guided by both digital and material sensitivity (Figliola and Battisti 2017). This type of knowledge is now teachable and transferable, supporting the development of a shared language between architects and designers, based on the definition and manipulation of design parameters.

Constructing Futures: The Fall of 2D Representation

In traditional construction, two-dimensional drawings serve as the primary means of communication between professionals, with digital technologies acting as "another pencil" (Scheer 2014). But what happens when digital information becomes completely self-sufficient, and data replaces drawings as the foundation for on-site execution? *Constructing Futures* is a built experiment that challenges and explores this question. The project was developed by Sara Codarin and Karl Daubmann as a part of the annual Detroit Month of Design event to provoke conversation around a speculative future of construction in the age of intelligent machines. The installation combined physical fabrication with a public exhibition of internationally curated, AI-generated images focused on the future of building (Fig. 2.8). These visuals served as a platform for public dialogue to focus upon emerging technologies, speculative practices, and future fictions (Bleecker 2022). Referencing an increasingly image-centric design culture (Young 2022), the project addresses how generative AI now enables the production of compelling, culturally relevant images worthy of critical engagement (Arieff 2025). The installation leverages low-budget off-site digital fabrication, on-site robotic layout, and off-the-shelf materials and hardware. Its material strategy reflects a circular, design-for-disassembly approach (Adekunle and Ben Alon 2023), making future repurposing and material recovery an integral part of the design logic.

The exhibition was located in Woodward 1001, a large gallery space in downtown Detroit. At its core is a curvilinear wood construction designed to capture the greatest volume with the least amount of material. The project occupies a 90-by-60-foot (27-by-18 meters) existing floor surface, shaping it into an inhabitable environment, framing a platform for dialogue through human-scaled experiences and the sensation of

Figure 2.8 *Constructing Futures*, a low-budget, digitally fabricated design-build exhibition framing a curated display of AI-generated images, 2024. Photo by Karl Daubmann.

being wrapped within a choreography of spatial moments. The tectonic system[15] of the full-scale installation is composed of surfaces treated as continuous elements that converge, diverge, and intersect to generate voided poche conditions (Fig. 2.9), drawing inspiration from the spatial sequencing of Baroque architecture, particularly Borromini's San Carlino alle Quattro Fontane (Mansure and Luarasi 2019). The design explores the concept of nested volumes, where surfaces simultaneously define interior spaces that are experienced from both within and outside of themselves. A single surface may operate on two sides—marking the interior of an outer shell or, conversely, the exterior of an inner void—composing a rhythmic progression of transitions (Fig. 2.10). These sculptural interstitial voids are brought into dialogue with thin plywood

Figure 2.9 Spatial poche volumes formed by continuous plywood surfaces supported by a timber framing system. Photo by Steve Kroodsma.

planes, which are selectively activated through various media. A curated selection of AI-generated images, arranged into a continuous ribbon, is integrated across the installation, unfolding along points of convergence, divergence, and cantilever (Fig. 2.11).

Budget constraints and a compressed construction timeline shaped the project's approach, prompting a fast, cost-effective strategy based on

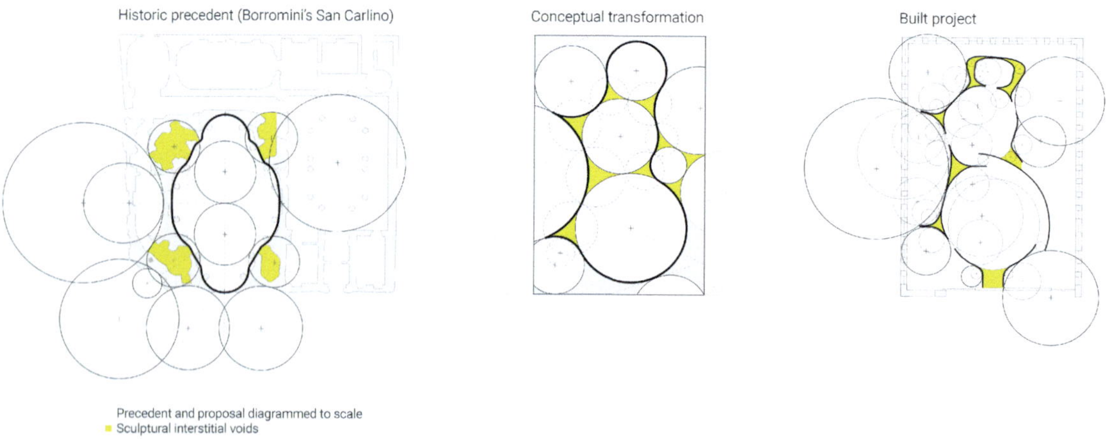

Figure 2.10 Concept diagrams. Generation of volumetric poche spaces that define spatial progression, inspired by the geometric relationships in Borromini's San Carlo alle Quattro Fontane.

Figure 2.11 CNC-fabricated, self-supporting stressed-skin structure shaping the installation's geometry, with layered surfaces generating in-between spatial conditions. Photos by Karl Daubmann (left), and Steve Kroodsma (center and right).

off-site prefabrication and the use of common, easily accessible materials (Schröpfer 2012). The vertical system consists of 2x4-inch (5x10 cm) dimensional lumber spaced at 16-inch intervals, clad with ¼-inch (6 mm) birch plywood to form the display surfaces. Stud walls are capped top and bottom with two layers of ¾-inch (2 cm) CNC-cut plywood. To streamline on-site coordination, the elevation was divided into buildable segments—0–4 feet, 4–8 feet, and 8–10 feet (0–1.2 meters, 1.2–2.4 meters, and 2.4–3.0 meter)—allowing components to be prefabricated and assembled independently. Maximum curvature was determined by the bending capacity of the plywood, achieving dramatic cantilevers while maintaining a rigid yet lightweight stressed-skin structure. The curved walls established a stable, free-standing geometry, eliminating the need to anchor any elements to the historic stone floor. Because construction was carried out by unskilled student labor, CNC components, such as top and bottom bands and stud placement markers, were designed to embed assembly intelligence directly into each part, simplifying and assisting the build process. To accelerate the layout implementation, an autonomous HP SitePrint robot was used to mark the floor plan and furniture positions directly onto the slab with temporary ink (Fig. 2.12). Reinforcing the project's focus on construction futures, a large-scale cartesian 3D concrete printer (Fig. 2.13) was also installed in the space and used to fabricate furniture during and after construction for use in the exhibition.

Figure 2.12 HP SitePrint utilized in the *Constructing Futures* design-build process, applying temporary ink to mark layout positions for component assembly.

Figure 2.13 Cartesian 3D concrete printer using off-the-shelf mortar to fabricate forms, serving as live R&D during the production of exhibition furniture.

Project constraints and the ambition to showcase advanced tools supported an efficient transition from design through installation. The entire fabrication and assembly process was completed in two weeks by a team of six individuals.[16] All components were derived directly from the 3D model, eliminating the need for a printed set of traditional construction drawings.

CNC-cut parts and the HP SitePrint robot ensured accurate placement and alignment on site. Throughout construction, the digital model was continuously referenced to verify field conditions and guide the process. An as-built LiDAR scan confirmed a tolerance within a quarter inch (6 mm) across the full 90-foot span (27 meters). At the end of the exhibition, the installation was fully demountable, with lumber and hardware preserved for future design-build projects. No on-site measurements were required; everything was laid out according to the model and robot-printed markings. Drawings were produced purely to meet permitting requirements and included just one layer of information: emergency exits and occupancy limits. As a small-scale prototype, *Constructing Futures* reflects broader possibilities and offers a perspective on how integrated workflows can help reduce fragmentation in design-build applications, highlighting the need for continued research to bring design, data, and execution into closer alignment.

Notes

1 The term Industry 4.0 was first introduced at the Hannover Messe in 2011 by Professor Wolfgang Wahlster, Director and CEO of the German Research Center for Artificial Intelligence, during the event's opening ceremony. He used the term in the context of preparing for success in a globally competitive, high-tech environment. Wahlster highlighted the need to be ready for the Fourth Industrial Revolution, driven by the rise of the Internet and digital technologies.

2 In today's digital age, connectivity offers new opportunities across the globe. However, even though mobile networks now reach 92 percent of the world's population, over 2.5 billion people remain without internet access. *The World Economic Forum*: https://www.weforum.org/stories/2024/09/2-5-billion-people-lack-internet-access-how-connectivity-can-unlock-their-potential/.

3 In *Graz Architecture Magazine*, Mario Carpo reflects on how the "old masters," associated with smokestacks, assembly lines, and strong trade unions, often saw the rise of nonstandard design as signaling the end of Fordism and, by extension, the weakening of labor unions. Many instinctively viewed digital technologies as a Trojan horse for neo-capitalism and neoliberalism, and thus met them with deep suspicion.

4 According to the Gartner dictionary, voxels (short *for* volumetric picture elements) are the 3D equivalent of pixels. In static volumetric displays, voxels are active elements distributed throughout a 3D space that can change color or transparency to form a visual representation of an object. See: https://www.gartner.com/en/information-technology/glossary/volumetric-displays.

5 Mario Carpo discusses the role of computers in the Second Digital Turn during the series Talks at Google in 2018.

6 "Scenarios, Pathways, and the Future-Ready Workforce." Article by Jim Phelps in *Educause*: https://er.educause.edu/articles/2018/8/scenarios-pathways-and-the-future-ready-workforce.

7 Digital Transformation is the Future of Higher Education. A roadmap by *Educause*: https://dx.educause.edu/.

8 "Digital Transformation Is Not About Technology". Article by Benham Tabrizi, Ed Lam, Kirk Girard, and Vernon Irvin in *Harvard Business Review*: https://hbr.org/2019/03/digital-transformation-is-not-about-technology.

9 Greg Lynn, How Calculus Is Changing Architecture, TED talk 2009: https://www.youtube.com/watch?v=DeyzUysMLy0.

10 In historic architecture, Palladio drew inspiration from shipbuilding; for example, in the construction of the roof of the Basilica in Piazza dei Signori, Vicenza, where he even employed craftsmen skilled in naval construction.

11 Ford's Willow Run Bomber Plant was designed by Albert Kahn and constructed in 1941. Photo documentation is available in the digital archives of the Henry Ford Collection: https://www.thehenryford.org/collections-and-research/digital-collections/.

12 An original printed copy of the drafting room at Ford Motor bomber factory (Willow Run, Michigan) by Hedrich Blessing Studio is kept in the Albert Kahn Library at Lawrence Technological University in Southfield.

13 In *Made by Robots*, Antoine Picon explains that Fuller saw the Dymaxion house, developed in the 1920s and 1930s, not just as an innovation in building, transportation, and everyday life, but as part of a broader vision for a radically different future in which individuals could move freely across the globe and fully exercise their intellectual capacities.

14 Spatial computing refers to technologies that enable computers to interact with the physical world in real time by integrating digital information with spatial, environmental, and sensor-based data. The term brings together extended reality (XR), the Internet of Things (IoT), robotics, and artificial intelligence, allowing users to visualize, simulate, and manipulate information within three-dimensional space.

15 A tectonic system refers to the integration of structure, materials, and construction methods into a cohesive and expressive architectural whole. Rather than focusing solely on functionality, this approach reveals the building's inner logic through its physical components, contributing to both its aesthetic presence and cultural meaning.

16 The *Constructing Futures* exhibition design was coordinated by Sara Codarin and Karl Daubmann, with construction and digital fabrication support from Sophia Grosan, Dane Peterson, Emmanuel Nava, Matthew DiMaggio, Reece Radomski, and Sarah Smith. Liaison with Bedrock Detroit, owner of the 1001 Woodward Avenue gallery, was facilitated by Ryan Mason and Matthew Churchwell.

References

Adekunle, Timothy, and Lola Ben Alon. 2023. "The Ingenuity of Circularity." *Technology|Architecture + Design* 7 (2): 181–82.

Aish, Robert. 2013. "First Build Your Tools." In *Inside Smartgeometry: Expanding the Architectural Possibilities of Computational Design*, Edited by Brady Peters, Terri Peters, 36–49.

Arieff, Allison. 2025. "AI Is Pushing the Limits of the Physical World." *MIT Technology Review*, April. https://www.technologyreview.com/2025/04/21/ai-is-pushing-the-limits-of-the-physical-world.

Bleecker, Julian. 2022. "Design Fiction: A Short Essay on Design, Science, Fact, and Fiction." In *Machine Learning and the City: Applications in Architecture and Urban Design*, Edited by Silvio Carta, 561–78.

Bonwetsch, Tobias, Fabio Gramazio, and Matthias Kohler. 2010. "Digitales Handwerk." In *GAM Architecture Magazine 06*, Edited by Martin Bechthold, Johan Bettum, Tobias Bonwetsch, et al., vol. 6. Graz Architektur Magazin Graz Architecture Magazine. Springer Vienna. https://doi.org/10.1007/978-3-211-99210-4_14.

Brennan, Annemarie. 2016. "Communicating with Machines: Bézier Curve as 'Informational Ontology.'" In *104th ACSA Annual Meeting Proceedings*. https://www.acsa-arch.org/chapter/communicating-withmachines-bezier-curve-asinformational-ontology/.

Brynjolfsson, Erik, and Andrew McAfee. 2014. *The Second Machine Age: Work, Progress, and Prosperity in a Time of Brilliant Technologies*. WW Norton & Company.

Carpo, Mario. 2013. *The Digital Turn in Architecture 1992-2012*. John Wiley & Sons.

Carpo, Mario. 2017. *The Second Digital Turn: Design Beyond Intelligence*. MIT Press.

Carpo, Mario. 2023a. "A Short but Believable History of the Digital Turn in Architecture." *E-Flux Architecture (March 2023). Https://Www.Eflux.Com/Architecture/Chronograms/528659/a-Short-but-Believable-History-Ofthe-Digital-Turn-in-Architecture*.

Carpo, Mario. 2023b. *Beyond Digital: Design and Automation at the End of Modernity*. MIT Press.

Figliola, Angelo. 2017. "Post-Industrial Robotics: Exploring Informed Architectures in the Post-Digital Era." *TECHNE – Journal of Technology for Architecture and Environment*, August 4, 256–66. https://doi.org/10.13128/TECHNE-19751.

Figliola, Angelo, and Alessandra Battisti. 2017. "Post-Industrial Robotics." *MD Journal*, 14, 14–25.

Germanà, Maria Luisa. 2019. "Technology and Architectural Heritage: Dynamic Connections." In *Conservation of Architectural Heritage*, Edited by Dean Hawkes, Hocine Bougdah, Federica Rosso, et al. Advances in Science, Technology & Innovation. Springer International Publishing. https://doi.org/10.1007/978-3-030-10871-7_7.

Kolarevic, Branko. 2004. *Architecture in the Digital Age: Design and Manufacturing*. Taylor & Francis.

Lynn, Greg, ed. 1999. *Animate Form*. Princeton Architectural Press.

Mansure, Adil, and Skender Luarasi. 2019. *Finding San Carlino: Collected Perspectives on the Geometry of the Baroque*. Routledge.

Murphy, Maurice, Eugene McGovern, and Sara Pavia. 2009. "Historic Building Information Modelling (HBIM)." *Structural Survey* 27 (4): 311–27.

Otero, Jorge. 2024. "Beyond Skills: Reflections on the Tacit Knowledge-Brain-Cognition Nexus on Heritage Conservators." *Heritage Science* 12 (1): 223.

Patankar, Yamini, Camilla Tennenini, Rafael Bischof, et al. 2024. "Heritage++, a Spatial Computing Approach to Heritage Conservation." *RILEM Technical Letters* 9: 50–60.

Pottmann, Helmut. 2013. "Architectural Geometry and Fabrication-Aware Design." *Nexus Network Journal* 15 (2): 195–208.

Quattrini, Ramona, Eva S. Malinverni, Paolo Clini, Romina Nespeca, and Erika Orlietti. 2015. "From TLS TO HBIM. High Quality Semantically-Aware 3D Modeling of Complex Architecture." *International Archives of the Photogrammetry, Remote Sensing & Spatial Information Sciences* XL-5/W4: 367–74.

Rifkin, Jeremy. 2014. *The Zero Marginal Cost Society: The Internet of Things, the Collaborative Commons, and the Eclipse of Capitalism*. 1st ed. Palgrave Macmillan.

Rudin, Rebecka, Malgorzata A. Zboinska, and Sanna Sämfors. 2023. "RePrint: Digital Workflow for Aesthetic Retrofitting of Deteriorated Architectural Elements with New Biomaterial Finishes." *Hybrids and Haecceities: Proceedings of the 42nd Annual Conference of the Association for Computer Aided Design in Architecture (ACADIA 2022)*, Philadelphia, PA, 336–45.

Scheer, David. 2014. *The Death of Drawing: Architecture in the Age of Simulation*. Routledge.

Schröpfer, Thomas. 2012. *Material Design: Informing Architecture by Materiality*. Walter de Gruyter.

Schwab, Klaus. 2017. *The Fourth Industrial Revolution*. Currency.

Sheil, R., Achim Menges, Ruairi Glynn, and Marilena Skavara. 2017. *Fabricate 2017*. UCL Press.

Tedeschi, Arturo. 2014. *AAD, Algorithms-Aided Design: Parametric Strategies Using Grasshopper*. Le penseur Publisher.

Young, Michael. 2022. *Reality Modeled after Images: Architecture and Aesthetics after the Digital Image*. Routledge. https://doi.org/10.4324/9781003149682.

Chapter 3

Industrialization of Building Products

DOI: 10.4324/9781003590873-3

A Conversation about Mass Customized Architecture

The interview that follows shares an insightful conversation between Karl Daubmann and myself about the present and future methodologies surrounding mass-customized architecture and construction.

Karl Daubmann is a valuable resource in the areas of design, industry, and academia, and his unique experience bridges the realms of mass production, prefabrication, and customization within the architectural design and construction industries. At the time of the interview, Karl is the Dean and Distinguished Professor[1] at the College of Architecture and Design at Lawrence Technological University and has an extensive background in design computing, digital fabrication, and multidisciplinary design. Karl served as the Vice President of Design and Creative Director for BLU Homes where he led multidisciplinary teams to develop innovative, customized, and prefabricated houses that were shipped across North America. He also co-founded PLY Architecture in Ann Arbor, Michigan, where he served as a partner for over ten years, focusing on projects where both design and digital fabrication were interwoven. Karl has a tacit understanding of the industry, and our interview discusses a shared interest in the future trajectories of the construction industry along with how to anticipate the changes that are approaching with the digital transformation that is upon us.

Sara: Thank you for taking the time to discuss these topics as a means of sharing your experience and framing relevant aspects between research and practice. To start, can you give me a bit of background on your work in industrialization and manufacturing?

Karl: My role as both a Dean and educator at LTU's College of Architecture and Design allows me to enjoy sharing stories and experiences from my practice with our students. However, practice or industry doesn't always map well onto academic contexts. My interests in design, technology, and manufacturing can be traced back through to my years of teaching, academic research, and design practice. I was a co-founder of PLY Architecture between 2000 and 2012 while also having the opportunity to teach several of these subject areas at the University of Michigan in Ann Arbor. In particular, our architectural practice at PLY completed many projects in this regard including a number of restaurant interiors where we designed and built a series of intense and custom installations. Given CAD/CAM technology, we were able to build these in quick and inexpensive ways because we leveraged the intelligence of our digital design models. In 2010, we were approached by BLU Homes to work as their creative direction team. This was an opportunity to develop houses as products and to be part of an integrated team that worked at a broader scale. Eventually, I left my practice and assumed the role of Vice President of Design at BLU Homes, where I oversaw product development, design, and collaboration in a greater capacity within the sales, marketing, engineering, and manufacturing teams. It was in this role that I began to understand the broader issues of complex, impactful problems and how design might have greater agency in these areas.

SARA: You mention being part of an integrated team at BLU Homes. One topic I'm particularly interested in is the notion of the master builder and the way that industry, education, and professional practice might evolve to address the increasingly complex world of construction. Specifically, when considering cultural heritage and its connection to the conservation of physical artifacts, changes in labor could affect design processes and the composition of on-site teams. What are your thoughts on the potential redefinition of the master builder role, and have you observed any evidence of this shift within the construction industry?

KARL: I've thought about this topic of master builder quite a bit and I think it emerges from my own experience because I grew up in a small fabrication shop and was exposed to construction and technology long before I attended architecture school. This experience at a young age informed the way I understood materials, innovation, and economics. In my own practice, I explored the potential of the master builder in two different ways, first as a boutique design-build practice, and second within an integrated team of both design and manufacture. At PLY we leveraged our knowledge of digital design and fabrication to handle all aspects of commercial interiors. We had control over materials, means and methods of construction, and understood first-hand how it impacted the budget. Initially, we were less confident because it was a non-standard form of practice, but as we advanced, we began to understand the contracts, roles, and economics in more sophisticated ways. In the United States, architects are screened from cost as this role is owned by the contractor, who holds more agency and risk in the process.

One thing that worries me about the trajectory of practice in the United States is the move to specialization. While specialization protects clients by giving the designer exclusive experience in specific building types, I see a lack of innovation occurring in these areas. Not all projects require creativity, but I fear that the rush to respond to liability risk may hinder progress. This relates back to the master builder concept because I envision a master builder being able to balance a deep an understanding of convention, innovation, and responsible cost management.

Designers in this type of role will give themselves incredible agency in the process while also balancing the relationship between risk and reward. I was speaking with Thom Leslie, Professor at the University of Illinois at Urbana-Champaign, a few years ago, and he explained to me that all of Pier Luigi Nervi's projects were competitively bid, meaning that they were cost-effective proposals. This blew my mind because we know Nervi's work to be beautiful and expressive, which often does not correspond to being the most cost-effective way to build. Nervi is a great role model as a master builder. Today, there is an opportunity to leverage the interconnected relationship between innovative design and construction, and the associated means and methods for its production. I don't think the master builder will be a singular person working in isolation but instead part of a collaborative and tightly integrated team. To successfully work in this manner, I believe it will also require changes to the way we educate young designers and builders.

I've always believed that our contracts, roles, and responsibilities define our professional actions. With BLU Homes, I was excited by the integrated model where the disciplines were all acting as one team within the same company. From the master builder's perspective, the design team played an important role as the integrator. Those of us responsible for product design worked with and supported the strategic and marketing teams. As project designers, we supported sales because we knew the products, codes, and the constraints. In addition, we fed accurate information to the engineering and manufacturing teams for the construction and permitting of the houses. We had designers, architects, programmers, BIM experts, technologists, and even a librarian on the design team. While I don't view any one individual as a master builder, the design team played a role like the one you suggest. In addition to your focus on the master builder being engaged in downstream construction activities, at BLU the design team had an equal effort upstream with aspects of sales, marketing, and client-facing aspects. In this way, I would argue that the master builder might be able to engage greater and greater amounts of data increasing the capability to implement a wide array of potential design inputs.

SARA: I want to pick up on your point about scaling things up. I am looking at the potential for robotics to be more widely deployed on cultural heritage sites which I plan to address. But before I do, I want to confirm that I have an understanding of industrialization and the mass-production landscape.

KARL: If I take a very broad view of industrialization, I am always amazed at the deeply held beliefs that reside in design and construction as it related to mass production. I began practicing in the late 1990s when mass-customization was a popular concept. As a frame of reference, prior to the Industrial Revolution in 1797 everything produced was a custom effort. Eli Whitney's patent for interchangeable parts on rifles initiates industrialization in the United States, but the proof of concept for the patent was actually three rifles, each made by hand to a high tolerance and the parts could only be interchanged upon those three guns. The challenge wasn't simply achieving standardization, but rather managing the costs associated with producing interchangeable parts and training a workforce skilled enough to consistently deliver such precision. Fast forward a century to 1926 and Henry Ford set records in Detroit, making 9,000 Model T vehicles per day out of his Highland Park Plant. This solidifies the belief that standardization makes things cheaper, which often it does until we dissect some of these parameters and constraints.

At one point, I had the opportunity to be part of the hiring process for a key position in manufacturing at BLU Homes. Through the interviews of the various applicants, I gained insight into a range of manufacturing strategies and the way they vary between industries and companies. I remember the story of an employee who was able to do a function in a fraction of the time that was specified. After further investigation, the approach was found not feasible to carry out by a variety of people nor sustainable for multiple workers. This is an example that supports the value of mass production biasing reproducibility and consistency over speed.

At BLU Homes, we seemed to always be pushing against the promise of the low cost of manufacturing. The mass production paradigm suggests that prices can drop with enough output, based upon the promise of mass production. However, at BLU, we never found this reduction through volume. The complexities came with what I would characterize as a much broader cultural issue of how people view their houses and how differently they operate when they feel responsible for designing them. There is a price point at which a client believes they should be able to customize the house, and we were building houses which were expensive by nature. Standardizing the houses meant we could do it faster and cheaper, but the market pushed against this, or more problematically, demanded the price of mass production with the flexibility of mass-customization.

From a manufacturing perspective, we tried both a high-tech strategy and a low-tech strategy. In both cases, a steel moment frame was the structural chassis that accepted variations in program configuration and enclosure, inclusive of solid walls, walls with windows, and large glass doors. In the high-tech approach, we were going directly from CATIA files to rolling our own light-gauge steel framing members, cut and punched, with zero waste. This required a mix of skilled labor and unskilled labor. But the market demanded reductions in cost and time, and BLU transitioned to a low-tech approach that consisted of simplified diagrammatic drawings for unskilled labor to work with traditional materials. This change required reworking products, engineering, drawing processes, and output to be able to be handed off to manufacturing. It resulted in a more rigorous process to produce a bill-of-materials for the purchasing department and also included the development of a database of all the required parts for assembly. This library of information was tied to the negotiation of contracts for purchasing materials and components in bulk. It was an incredible process to be part of, as I watched the changes in the design and collaboration process be driven by what I soon realized was a complex manufacturing ecosystem with every aspect in full alignment. In the end, I don't believe the low-tech system was able to deliver the desired price or time reductions.

To this day, I still consider the massive upfront investment required for manufacturing our process efficiently, and I don't know if the long-term costs outperformed the mass-crafted approach that one finds in architecture where each house is a unique prototype. Over time with enough houses built, this likely reaches a tipping point, but at an early stage of the endeavor, we were treating the houses as products long before the entire system had been prototyped.

While the manufacturing was being tweaked, the design team was also tasked with retooling the products to be more manufacturable. One example would be a standard set of window sizes that until then had been based upon available sizes from window manufacturers and a model-by-mode or case-by-case basis. The reduction in window types went from about 50 types to more like 10 window types. This reduced the amount that the factory had to manage and allowed storage of windows and bulk ordering discounts that were not feasible previously. Beyond design for manufacturability, the design team also needed to confront the high number of customizations requested by clients. To address the constraints of flexibility and price, we developed packages of finishes and options, mimicking that of the automotive industry. We could pre-design these offerings to make sure they worked with manufacturing and confirm that they could be supported by market research. As a result, our digital parametric models were tied to a client-facing, online configuration tool. With this web-based tool, not only could a client make selections of house plans, kitchen types, and finish and fixture choices, but they could also walk through a virtual version of the house and see the fixed price of their decisions.

SARA: How do you envision this type of work moving from mass production to mass customization? Architecture is inherently a small run, one-off, mass-crafted approach. Heritage sites are a more extreme version of one-offs and would benefit significantly if being able to leverage mass-customization.

KARL: The gap between design and construction seems to be closing, in a large part because of the ability for designers' software to communicate with fabricators' hardware. Therefore, I believe that the move to mass customization occurs first with design tools. Parametric and algorithmic software help us conceive things more as a system instead of as individual one-offs. If designers can also understand manufacturing and construction processes more fully, we can find opportunities for variations. That deep construction knowledge could lead to mass customization.

SARA: Part of what I am investigating within my research is the implication of technology on the design process. Can you speak to the workflow from design to manufacturing as it relates to the construction process in the United States through your experience at BLU Homes?

KARL: I often reference a simple diagram from Marty Neumeier's book *Metaskills* that illustrates automation on a spectrum (Neumeier 2013). At BLU we developed plug-ins within our Revit software to automate many of the tasks that were cumbersome and had accuracy issues. We looked at it from both the perspective of the designer and the client, then automated the required output for the factory. This process was incredibly effective but only worked in instances where the process was repetitive. We were customizing products. If one understands the scale of repetition, it could be deployed in different ways. We already see this in relation to parametric design and BIM where many drawings are extracted in an automated way. We automated the floor and wall framing members and their spacing where objects placed by designers would either attract or repel framing members. For example, if a wall switch was located, it required a vertical stud to be placed adjacent to it, whereas a toilet would require that floor framing members avoid the toilet fixture location. Onsite, carpenters can look at plans and make adjustments. However, in the factory setting the responsibilities were separated by manufacturing stations, which required only the framing station having to know where the structural members were placed. This is an example of the infrastructure behind the scenes being customized based on client choices. Through automation, we were able to make these customizations without any schedule or budget implications. It's minor but exemplifies small-scale mass customization from my perspective.

SARA: Speaking of process innovation, the first industrial robot was used in 1969 at a General Motors plant. Yet, even after fifty years, we still haven't witnessed a large-scale transfer of this technology into the construction industry. Do you have a sense of why this might be?

KARL: I would consider at least five reasons why this would be true. First, robots have been expensive but the prices are decreasing. Second, the required technical knowledge to run robots has been a specialty. Third, robots have been used predominantly in repetitive tasks whereas construction seems to be structured in a different way. Fourth, the construction industry is fragmented and finally, generally a construction site is not currently set up to accommodate the needs of industrial robots.

Thankfully, there are responses to all of these concerns. LTU has been working with Ballard International, which is a company founded by LTU alumnus Justin Ballard, and specializes in buying and selling used robots and robotic systems. I see Ballard serving as a unique inflection point in the Fourth Industrial Revolution. They buy robots wholesale from the automotive industry, ensure their quality and function, and resell them to different industries that can access these technologies with less investment. People are coming out of school with more of this knowledge as digital fabrication has recently embraced robotic fabrication. In the same way that digital fabrication led the change from mass production to mass customization, the use of robots for prototyping will make them more adapt for non-routine tasks. If sensors are introduced on the robots, it will be even better.

According to the American Institute of Architects (Baker and Riskus 2008) and National Council of Architectural Registration Boards 2024 (NCARB 2024) annual reports, there are currently over 121,000 licensed architects and 19,000 architecture firms in the United States. This means there is very little consensus and minimal venture capital because of a distributed and fragmented market. According to the Association of General Contractors of America,[2] while the construction industry has a high revenue, it is split among almost 943,000 construction establishments nationally. This means that if someone wants to revolutionize the industry, it's difficult to do because it is incredibly localized. In industry, much of the automation is for simple or singular purposes, where you can justify automation advancement of one step in the process. Robots are interesting because of how flexible they are, but they are also a paradox because it hasn't been proven exactly how well or accurately, they could solve one dedicated problem on a construction site.

Sara: In the 1970s, some big construction companies experimented with single tasking robots onsite but it seems that this didn't continue. In addition, robot programming was used to differentiate the tasks that a single robot could perform. This was a way to pioneer the customization of production, and is still an ongoing conversation today. What are the main drivers that stopped that disruption in construction from happening? Are we at a changing point now?

KARL: Construction is in the business of customization. Every building site is unique. Every structure built has some level of uniqueness. Given this, the industry is set up in an individualistic basis. Actually, I think this is problematic because I'm not sure that the automation disruption in construction ever started. The industry is set up to deal with individual clients, many of whom have never been a client before. Every building constructed reflects individual program requirements. Our industry suffers because it is common for buildings to be over budget and behind schedule, and mass production has been able to deliver in both of these arenas. This puts the perception of our industry at a disadvantage. The question could be: can we deliver customization with reliability? Means and methods are conventional but their deployment seems to vary. What are the construction tasks that might be turned into standard routines as a means of increasing our performance and reliability? Architects have long complained about the quality that results from the construction industry and now thanks to digital design, one could imagine that the construction process could be pre-simulated. Architects might be in a great position to introduce robots into the process because of the design tools at our disposal.

SARA: I came across information from euRobotics that addresses a roadmap for the future of European robotics in 2030.[3] The roadmap utilizes a multi-layered approach and is intended as a living work document that foresees future developments and timelines for the application of robotic technologies in different sectors.

KARL: Which sectors do they highlight?

SARA: Medicine, surgery, agriculture, and assembly. Is there a chance that designers could be early adopters of these technologies?

KARL: Sadly, probably not. Which is only to say that since culturally we value the longevity of buildings, contractors bear the risk if a building does not perform. This means that the construction industry will probably never be the early adopter that you hope for it to be. But in the digital fabrication course that I taught, I loved to emphasize the connections between Frank Gehry's designs, from the furniture to the bus stop, to the Fish sculpture, and finally to the buildings. Differently scaled projects allow for various levels of risk and innovation. I used digital fabrication in many restaurant interior projects that never had to deal with enclosures and allowed for investigating ideas about non-standard parts, while not having to worry about waterproofing.

SARA: One last question, did BLU Homes consider using robots?

KARL: This is a very good and complicated question. First, I would say that I finished working with Blu Homes in 2014. At that point I went back to teaching full-time and one of my courses was an introduction to robotic fabrication. So, I have to admit that I didn't have robots on my brain until that point in time. However, Blu Homes did experiment with digital fabrication tools and file-to-factory processes at an early stage in its development. Those technologies had to do more with components and sub-assemblies than the actual assembly itself. While I could imagine some of the work in the factory being carried out by robots, I think the economics might have been the hurdle. We located our factory in a geographic location where there was a large pool of motivated and experienced workers. The large size of the pool meant that the cost of automating would always be higher. Your question also leads to the acknowledgement that geography, time, and many external factors might govern when it is appropriate or cost-effective to deploy robots. I know of many markets where labor is scarce and robots may serve a need sooner. For example, a unique Dairy Farm in Northern Michigan uses robots to milk their cows, mainly because they can't find anyone willing to do that work. We will continue to watch for industries where robots pop up and it's likely determined by the complexity of this type of equation.

SARA: I'm incredibly interested in this topic because I believe that robots are an important next step in the evolution of cultural heritage due to the restoration market being both a premium skilled labor and an aging population of craftspeople. Thanks for taking the time to share your insight about these issues.

Pioneering On-Site Building Automation

Since the early days of assembly lines, numerous attempts have been made to automate housing construction (Keating et al. 2014). One of the earliest examples dates back to 1917, when Thomas Edison patented a single-pour concrete system for building standardized houses by using a reusable mold. This concept anticipated the industrialized construction processes of the 1920s and 1930s, when the serial prefabrication of building components evolved into the mass production of standardized homes. Industrialization influenced the work of modernist architects, forcing designers to think differently about the aesthetic values of buildings as products. Advances in machinery, serially produced components, and industrially fabricated materials made construction more feasible and efficient. In the 1920s, minimalism emerged from a context marked by the duality of cheap materials and expensive labor. By contrast, the postwar period saw a global reversal: material costs began to outweigh those of labor. The rise of the Second Digital Age introduced digital fabrication with new formal possibilities and a renewed interest in compositional and decorative complexity.[4] As skilled labor becomes increasingly scarce and costly in the construction sector, the declining cost of automated industrial robots continues to reshape the landscape of architectural production (Fig. 3.1).

The introduction of automated digital technologies on construction sites in the 1960s and 1970s reflected a vision of modernity, using technology to manage dimensional tolerances and improve precision. During this period, on-site robotics were also developed to streamline construction by reducing the number of stages, leveraging unskilled labor, and accelerating project timelines (Bock and Langenberg 2014). The origins of building site automation, however, trace back to the production process transformations of the First Industrial Revolution. Infrastructure expansion

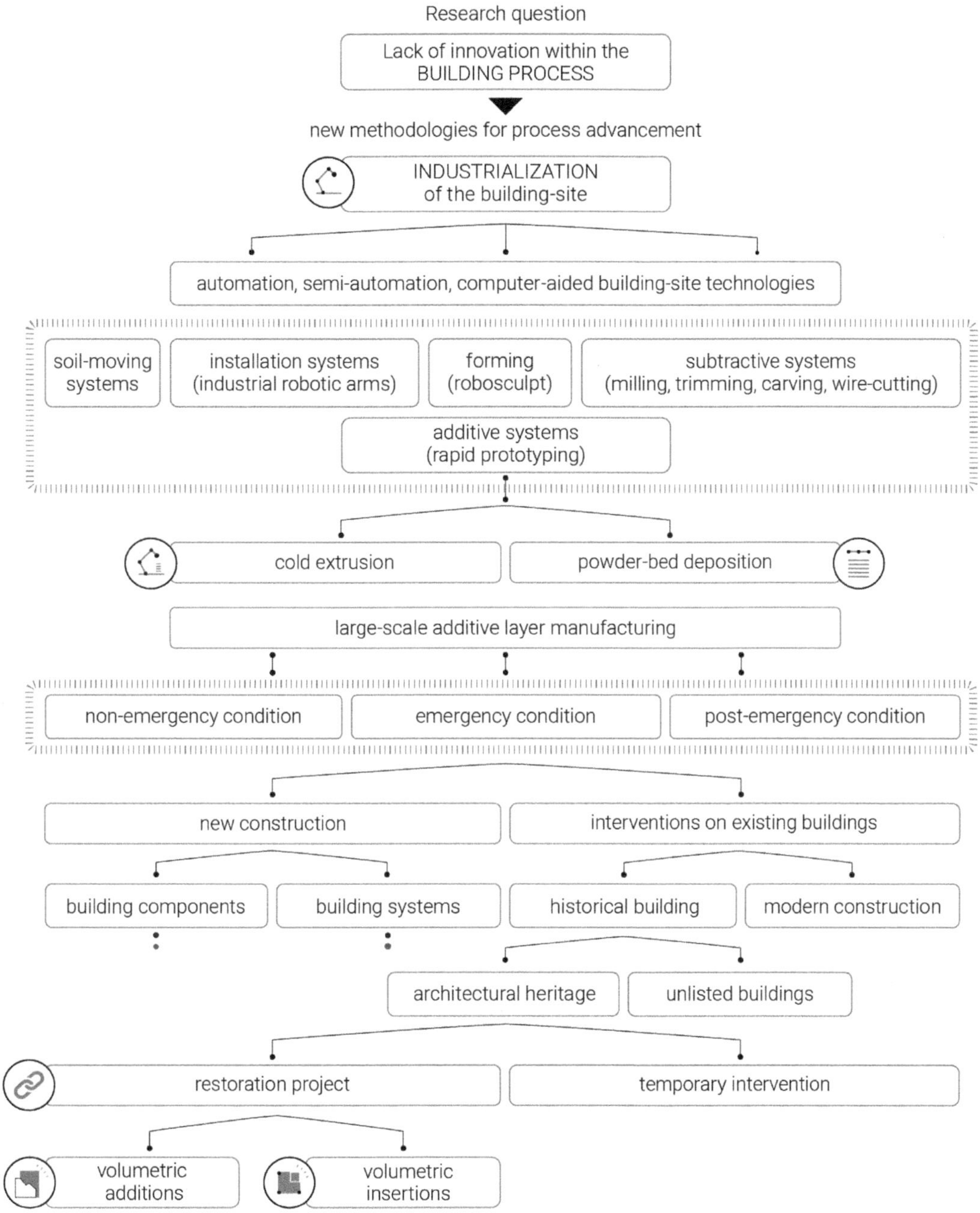

Figure 3.1 Envisioning the industrialization of the restoration building site, highlighting how tools and technologies from the Fourth Industrial Revolution can optimize the production and deployment of building components.

facilitated greater mobility between countries that supplied industrially produced materials such as cast-iron beams and glass, along with skilled labor, construction knowledge, and local building traditions. The Crystal Palace constructed for the World Exhibition in 1851 in London symbolized this transformation, presenting the latest industrial achievements from around the world and expressing the era's optimism toward innovation and technological progress (Nardi 1980).

Throughout the 20th century, avant-garde architects proposed radical alternatives to conventional building practices that anticipated the future of industrialized construction. Buckminster Fuller's Dymaxion House,[5] completed in 1930, introduced pioneering design principles alongside exploratory construction techniques and materials. In the 1940s, Jean Prouvé developed flat-pack, lightweight housing prototypes designed for global distribution and rapid on-site assembly. Around the same time, in 1942, Konrad Wachsmann and Walter Gropius designed the Packaged House System, a modular prefabrication construction method that embraced standardization and rapid assembly. On a larger scale, the United States launched Operation Breakthrough in 1968, aiming to provide housing for 26 million families over a decade, an ambitious goal that ultimately remained unfulfilled.

In Italy, the post-war period of the 1950s and 1960s marked a growing interest in standardization and modularity within the construction sector. Professionals increasingly focused on unifying construction materials and developing projects that were planned in full technical detail from the outset (Arbizzani 2015). The aim was to extend industrial production methodologies to architecture through what became known as closed prefabrication, a system that emphasized rigid serial production to deliver cost-effective and predictable technological units. This approach relied on minimizing the range of materials used, allowing components to be assembled and installed by predominantly unskilled workforce. As a counterpoint, attention gradually shifted from the full industrialization of entire projects to the refinement of techniques for on-site construction of individual building parts. This transition led to a renewed understanding of the building system and introduced the concept of flexibility, increasingly associated with the emerging approach of "open prefabrication" (Arbizzani 2015). Open prefabrication made use of standardized, catalog-based components designed for compatibility and quick assembly, while also supporting production variation and giving designers greater freedom to adapt and customize their solutions.

During the 1980s, Italy saw a surge in theoretical frameworks related to construction processes. In *Progettare nel Processo Edilizio*, edited by Mario Zaffagnini, Pierluigi Spadolini proposed a systematization of building construction by classifying the production of technological units into three categories: industrial, semi-industrial, and craft-based (Spadolini 1981). In the same publication, Ivan Cicconi outlined two primary approaches to building industrialization: prefabrication and on-site industrialization. Prefabrication referred to the off-site production of building components, while on-site industrialization was based on the direct fabrication of units on the construction site (Cicconi 1981). Cicconi further distinguished between heavy and light prefabrication, depending on the materials used. Heavy prefabrication typically employed traditional materials such as concrete and aggregates, whereas light prefabrication utilized industrially derived materials like steel and petroleum-based products. Amid the wave of experimentation of the 1980s, the pursuit of innovation led to the adoption of climbing formwork, a hydraulic system that enabled the in-situ industrialization of high-rise buildings and automated several construction phases. The first comprehensive application of this technology in Italy was led by Mario Zaffagnini, whose work introduced critical questions into architectural discourse regarding the limitations such systems might impose on design expressiveness and spatial flexibility. Despite its significance as a technological and theoretical benchmark, this advancement was not followed by further disruptive innovations in the field, largely due to the absence of a coordinated agenda and insufficient institutional support.

A broader international shift also influenced the trajectory of building automation during the 1970s. Contrasting economic conditions in Japan and the United States shaped divergent approaches to construction automation. In the US, the 1973 oil embargo triggered a recession and brought awareness of the limits to economic growth. As a result, efforts to fully industrialize building processes declined or were abandoned. Meanwhile, in Japan, a growing population created increased demand for social housing. The shortage of skilled labor in the construction sector led to the pursuit of alternatives to traditional building methods, encouraging the development of automation in both prefabrication and on-site construction. As Thomas Bock and Thomas Linner explain in *Changing Building Sites: Industrialization and Automation of the Building Process*, Japan invested heavily in research focused on automated housing prefabrication, single-task construction robots for on-site applications, and the development of fully integrated automated construction sites. Japan shifted construction activity from the building site to a more structured, factory-based environment, where 85 percent of the work was carried out off-site by human labor. Unlike European models of prefabrication, largely focused on producing large quantities of identical, low-cost components, Japanese prefabrication was oriented toward customization and adaptability. The combination of assembly line efficiency with the flexibility of manual labor allowed individual parts to be removed, modified, and reintegrated into the production process without disrupting the overall workflow (Bock and Linner 2016). This ability to customize components while preserving productivity laid the foundation for the contemporary integration of robotics in architecture.

The advancement of research in this field during the 1980s was fueled by the "robot boom" that emerged in the manufacturing industry. As a result, the adoption of robotics became a logical progression for Japanese construction firms, leading to the development of single-task construction robots. These robots were designed to perform specific tasks repetitively and proved valuable across a wide range of construction activities, including demolition, surveying, excavation, paving, tunneling, transport and distribution, casting and finishing, welding, steel positioning, fireproofing, paint spraying, inspection, and maintenance. Construction sites began to be structured and managed similarly to factories, with the ultimate goal of implementing fully automated manufacturing and construction systems.[6] A reorientation occurred from single-task construction robots toward integrated, automated construction for larger structured environments.[7] Advanced construction sites began to function as vertically mobile on-site factories, offering a controlled and systematized workspace. These platforms provided shelter for both assembly and disassembly processes, protecting operations from weather conditions and enabling the efficient handling of prefabricated, modular components with varying levels of detail. The shift toward integrated automation supported the coordinated deployment of robotic technologies through the integration of automated cranes, computer vision systems, and real-time control mechanisms. In this context, the earlier concept of climbing evolved into what became known as sky factory,[8] an automated or robotic on-site system designed specifically for the "manufacturing of vertically oriented buildings" (Bock and Linner 2016). The sky factory allowed the workspace to rise in tandem with the construction of each floor, enabling the precise positioning of building units using robotic manipulators and actuators. Japanese contractors also developed partially automated deconstruction systems, applying the same principles in reverse to reduce noise, dust, and disruption to the surrounding environment.

Despite these advancements, the broader effort to automate construction ultimately faced scalability challenges. One major barrier was the high upfront cost of robotic systems, which restricted their application to specific contexts such as urban areas with higher labor costs. Technical limitations also played a role, as the available computing power at the time was insufficient to fully support complex automation processes (Fig. 3.2). In addition, the absence of clear regulatory frameworks further slowed development and adoption. These obstacles encountered with early automation efforts revealed that large-scale integration of robotics in architecture demands more than the replication of existing construction methods. Successful implementation requires rethinking the design process from its early stages, along with adapting construction workflows to accommodate new technologies.

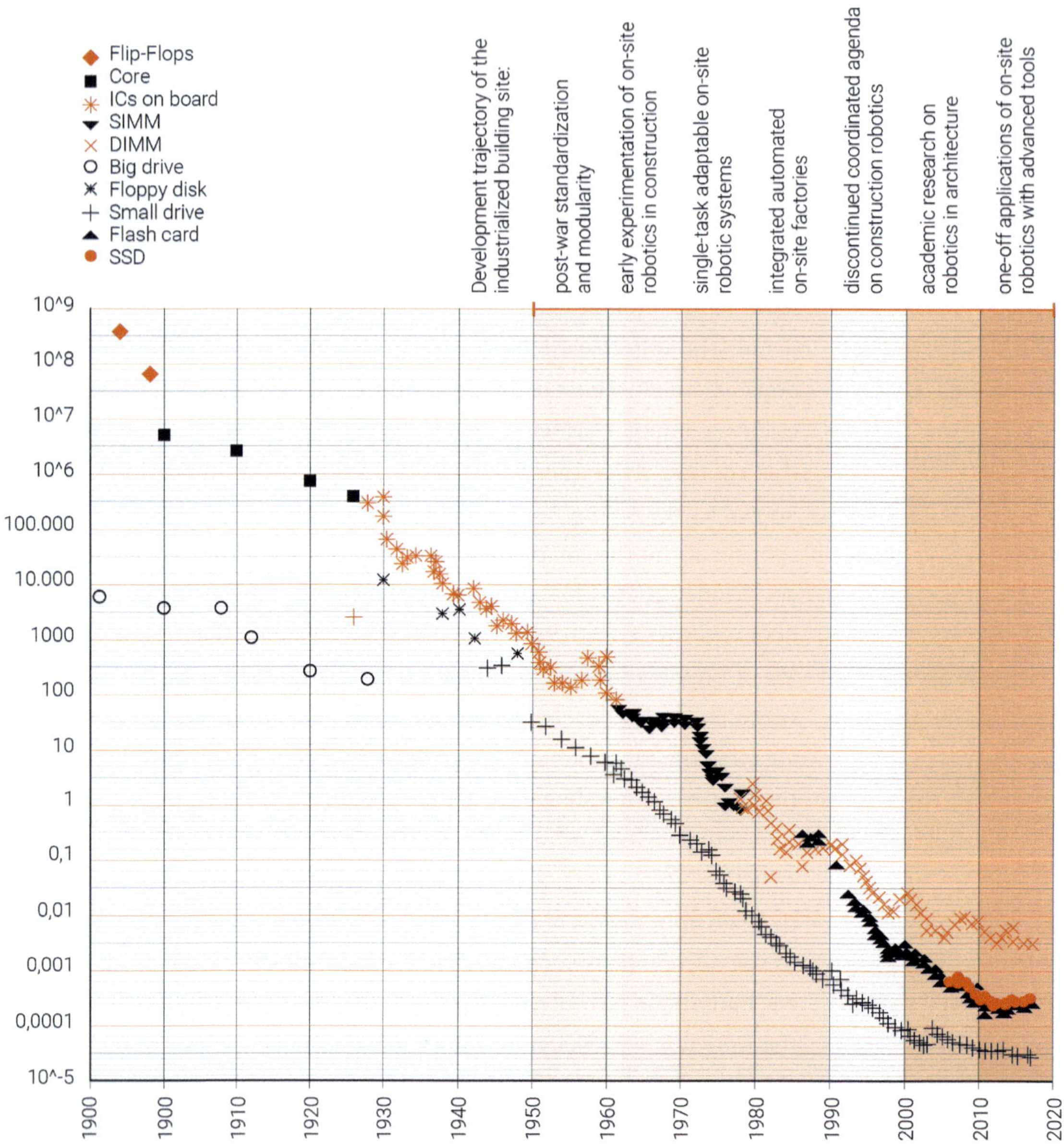

Figure 3.2 Historical cost of computer memory and storage. Between the 1960s and 1980s, the cost of 1 MB averaged around $50, while 1 GB was valued at approximately $50,000. Diagram adapted from data compiled by McCallum and Block, licensed under CC BY-SA.

Types of Contemporary Automation

Advancements in automated construction are steadily progressing, with flexible industrial robots now widely adopted in both research and practice (Petersen et al. 2019). Ongoing investigations explore several key strategies for construction robotics, including large-scale automated additive systems, cable-suspended platforms for dynamic positioning, swarm-based approaches modeled on collective behavior, versatile industrial robotic arms capable of executing a range of tasks, mobile robotic units designed to navigate complex site conditions, and folding systems that manipulate materials for efficient assembly and deployment.

An early and influential concept in large-scale additive manufacturing (AM), from which multiple systems have evolved, is Contour Crafting, developed by Behrokh Khoshnevis at the University of Southern California. This method pioneered a scheme for deploying automated construction using a bridge-like gantry that moves horizontally along two parallel tracks, guiding an actuator whose motion is controlled by cartesian coordinates. The process supports on-site construction of entire buildings through the layer-by-layer extrusion of full-scale components (Khoshnevis 2004). Cable-suspended platforms represent another approach, consisting of automated technologies in which multiple cables control the movement of a mobile. This method functions similarly to a highly precise crane and delivers a high degree of control and flexibility (Dubor et al. 2019; Sousa et al. 2016). The Italian company WASP employs large-scale AM systems, including cable-driven platforms, to construct full-scale buildings using natural materials. In such systems, the movement of the end-effector is controlled along dynamic paths, allowing for continuous and adaptive construction processes. The swarm-based configuration relies on a network of small autonomous units programmed to work cooperatively toward a shared objective. This method often incorporates flying autonomous vehicles capable of manipulating building components mid-air, reducing dependence on ground-based mobility and eliminating the need for cranes (Willmann et al. 2012; Zhang et al. 2022). Multipurpose robotics, often implemented through industrial robotic arms, are adaptable machines capable of performing autonomous tasks, with kinematics defined by rotational movements along multiple axes, typically within a polar coordinate system. They operate through actuators or end-effectors programmed to follow additive, subtractive, smart assembly, or hybrid fabrication logic. Mobile robotic units fall into a more dynamic category, as they are designed to move around and grant access to work areas unreachable by fixed setups. Examples are the prototype *Spot* by Boston Dynamics, designed to monitor building site conditions, and the mobile robotic platform developed at ETH Zurich for the construction of the fully digitally fabricated DFAB House (Dörfler et al. 2019). Distinct from the systems above, folding setups are designed to reshape construction materials at an intermediate, architectural sub-system scale, supporting the fabrication of structural elements that bridge the gap between individual components and the larger coordinated frameworks they reinforce.

In parallel with the development of these automated strategies for new construction, research practices are also examining how automation might support circular design approaches, particularly in the reuse and reprocessing of architectural components. The project *Designing Out Waste*[9] investigated the multi-scalar nature of waste, including urban, material, spatial, and energy waste, and proposed "robotic refabrication" (Jensen et al. 2019) as a method for re-manufacturing and re-purposing elements from existing buildings. This process envisions off-site transformation facilities that operate as fully automated lines for processing salvaged materials, embracing principles of circularity and design for disassembly (Fig. 3.3).

The integration of robotic arms, automated machines, bots, and drones into construction workflows expands the boundaries of digital crafting by enabling these technologies to take on tasks traditionally performed by human laborers. When guided by human design intent, these machines can support and enhance nearly every phase of the fabrication process. Their use helps streamline production by saving time and reducing costs, while also making it easier to explore new design qualities

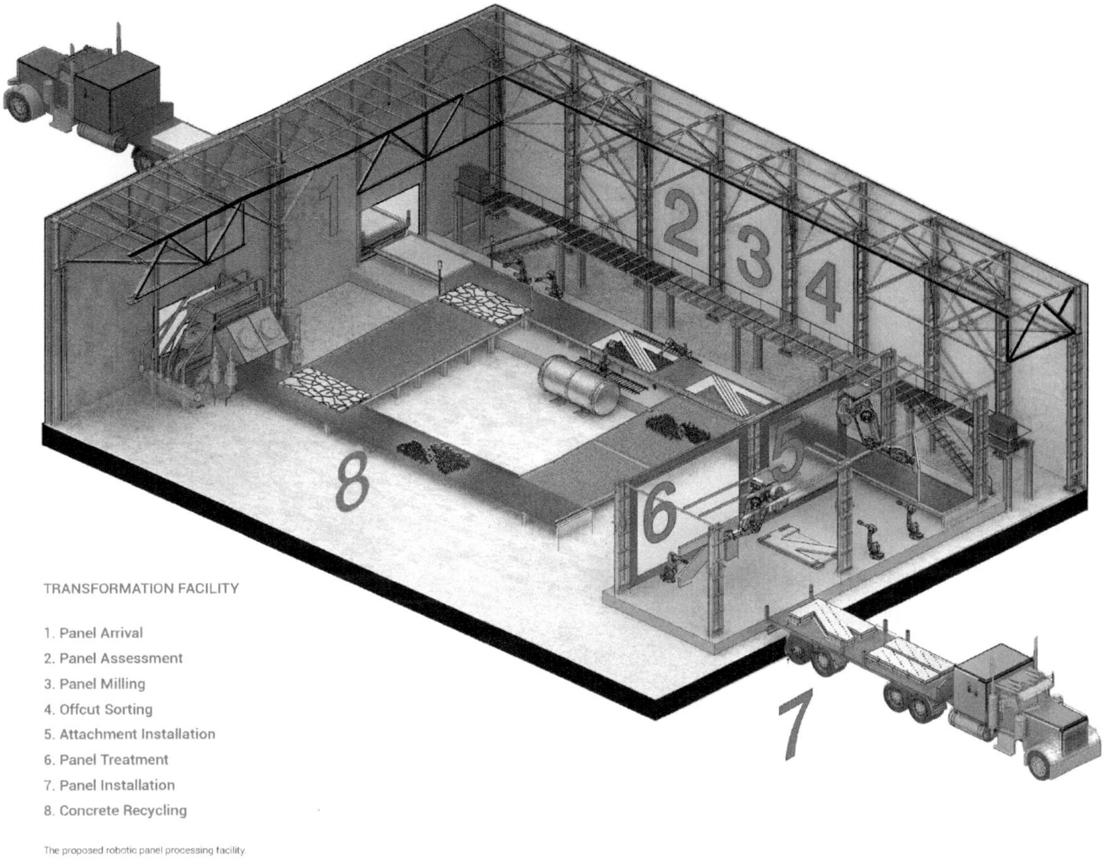

Figure 3.3 Proposal for a robotic transformation facility illustrating the off-site refabrication process of reclaimed facade panels through assessment, milling, treatment, and reintegration. Source: *Design Out Waste. A Brief on Robotic Refabrication* by Kasper Guldager, Joshua Taron, Kristen Forward, and Hayden Pattullo.

and fabrication methods. As part of custom digital workflows, these tools support more thoughtful ways of working with materials, enabling the upcycling of both common and newer types. Together, these capabilities contribute to a more flexible and resource-aware approach to building.

Additive Manufacturing

AM, or additive layer manufacturing (ALM), commonly referred to as 3D printing, emerged in the 1980s, when Hideo Kodama of the Nagoya Municipal Industrial Research Institute filed a patent application for a rapid prototyping technique. The first AM system to be introduced on the market was stereolithography,[10] officially launched in 1986 by Chuck Hull, founder of the company 3D Systems Corp. According to the *Gartner's IT Glossary*, AM is the capability to produce a physical object from a digitally encoded design through material deposition using a 3D printing process. The American Society for Testing and Materials (ASTM) defines AM as a process of joining materials to produce parts from 3D model data, typically layer by layer, in contrast to subtractive and formative manufacturing methods. AM is commonly employed as a rapid prototyping technique, facilitating the validation of digital design iterations through material-based processes. This process follows a logic of creating custom parts in an economy without scale, as it fundamentally differs from mass-production methods. Traditional manufacturing relies on molds and dies to shape raw material, then removes excess by cutting and grinding from a solid block to produce a repeatable result.

Large-scale AM typically refers to the creation of building elements at the architectural scale (Labonnote et al. 2016), enabling either custom prefabricated assembly systems or direct on-site printing. Scale is a defining factor, as additive construction represents the entire process of building through programmed automated machinery. The process relies on the deposition of material layers based on a digital model, following a file-to-production workflow capable of reading digital data, interpreting spatial characteristics, and opening up new languages that require technical experimentation and testing applicability. 3D printers are designed to support the leap in scale from small-scale design objects to full-scale building components, helping to overcome the constraints imposed by traditional construction methods.

The progression of AM in architectural research is well documented by organizations like Association for Computer-Aided Design in Architecture (ACADIA). ACADIA's conference proceedings archive, dating back to 1985, first mentions large-scale manufacturing in a 2009 contribution that examines 3D printing engineering technologies capable of producing construction components and architectural elements (De Kestelier and

Buswell 2009). The article also examines various production systems, such as contour crafting, powder-bed deposition, and robotic extrusion of fast-setting aggregate-free mortar, also known as concrete 3D printing, as drivers for unprecedented geometric freedom. In the 2010s, new methods have emerged, such as WASP Delta technology, Apis Cor's mechanical manipulators, and drones programmed to deposit cement polymers. Other experiments have investigated the use of materials like raw earth, cellulose-based biopolymers, hydrogels, and multi-material combinations. Nevertheless, the construction industry still predominantly relies on mortars and cement mixtures due to their widespread availability and ease of certification, as shown by projects from ICON Build, CREATE at the University of Southern Denmark, Vertico, XtreeE, and TwenteAM.

AM technologies can generally be classified into three types: powder-based, liquid-based, and solid-based systems. Processes are also classified by the type of material used or the technical characteristics of the fabrication machines involved (Fig. 3.4). AM was mainly used for rapid prototyping in its early stages, whereas now large-scale 3D printing has expanded its field of investigation for product innovation. To generalize, the AM workflow starts by decomposing the design's poly-surfaces into horizontal layers and dividing them into target points, converting pixels into voxels. To ensure print stability, parameters such as layer thickness, intra-layer metal reinforcement placement, extrusion speed, and material setting times are adjusted accordingly.

AM can be applied in the architecture, engineering, and construction sectors in two principal ways: either by producing components and subcomponents for later assembly into larger structures, or by directly printing large-scale, self-supporting architectural elements in their entirety (Naboni and Paoletti 2015). Although it is not yet cost-effective and still regarded as a premium technology, AM continues to attract strong research and development interest due to several promising advantages. These advantages include placing material only where geometrically necessary which reduces resource consumption and waste; the ability to create complex morphologies; the emergence of a distinct digital aesthetic;

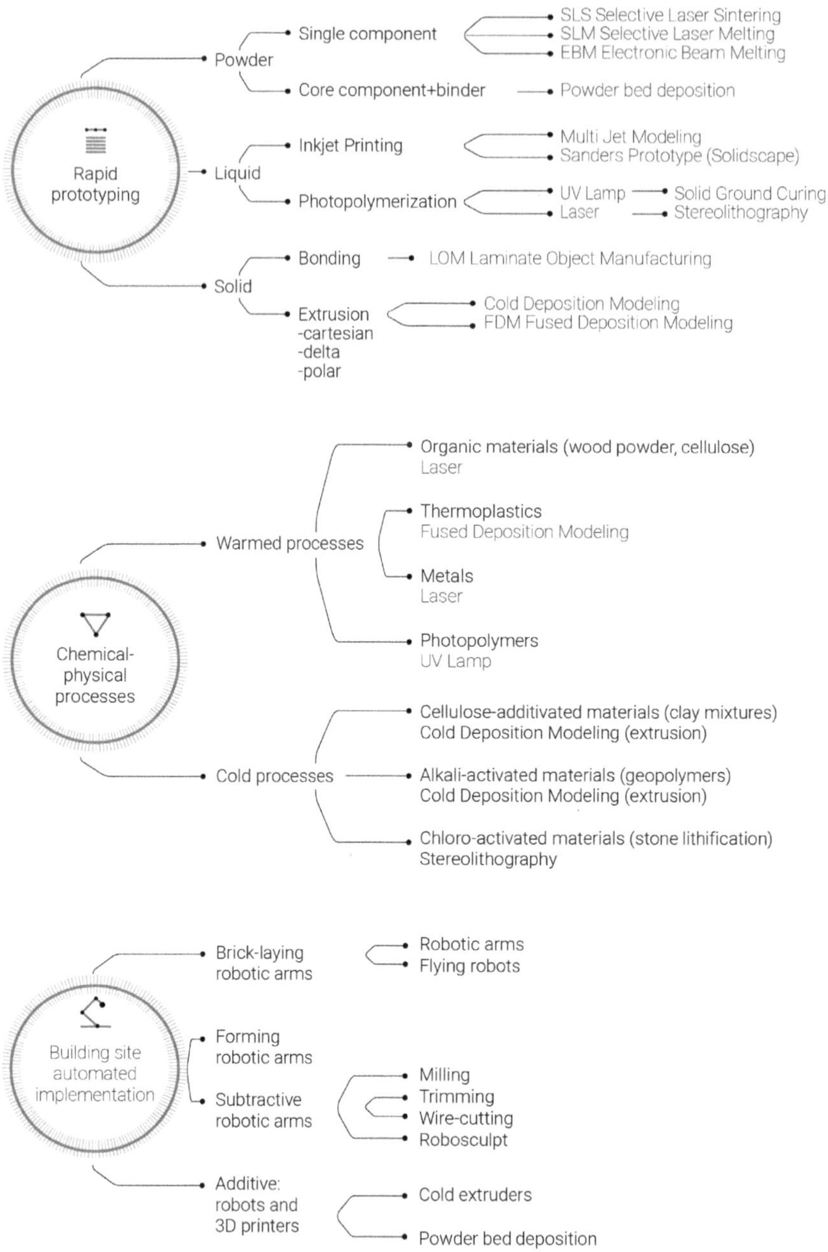

Figure 3.4 Classification of rapid prototyping technologies and their applications. The diagram outlines materials suitable for additive layer manufacturing and the corresponding types of machinery used in these processes.

extended design timelines that support optimization and predictive modeling; simplified installation and assembly of components; precise control over dimensional tolerances; and reduced reliance on conventional formwork or molds. While many projects have adopted the contour crafting method, where the building takes shape within the fixed printing area of the machine, this spatial constraint remains a limitation that researchers aim to overcome. Another major challenge is that most printed elements are not load-bearing, which complicates the creation of horizontal structures such as floors. Additionally, there is a general lack of widespread expertise, along with the absence of a standardized regulatory framework for digital manufacturing workflows. The potential to shorten the construction supply chain by producing components off-site is often cited as a major benefit, yet in practice, it remains costly and demands considerable logistical effort for transportation.

In the meantime, AM continues to advance, with two leading technologies driving innovation at the architectural scale: powder-bed deposition, based on matrix-oriented programming, and continuous extrusion, guided by toolpath-based control. Powder-bed deposition systems operate by layering a base material, typically a fine inert substance such as sand or gypsum, with an inorganic binder that solidifies the aggregates, applied through alternating nozzle passes. These systems enable the creation of monolithic, free-form volumes without geometric constraints along any axis (De Kestelier 2011). In 2007, the Italian company D-Shape, in collaboration with Shiro Studio, used this technology to produce *Radiolaria* (Fig. 3.5), the first large-scale digitally fabricated prototype analogous to a building unit (Morgante 2011). The design was based on algorithmic rules derived from the geometry of radiolarians, microscopic marine organisms ranging in size from one-tenth to one-hundredth of a millimeter. Their unicellular structures exemplify morphological performance optimization, serving as a biological model for architectural form. This biomimetic approach made AM the logical construction method, as it allows material to be deposited only where needed, minimizing waste, much like the growth processes observed in nature (Menges 2012).

Figure 3.5 Andrea Morgante (Shiro Studio) and Enrico Dini (D-Shape) with *Radiolaria*, the first large-scale 3D printed prototype made with a monolithic stone-like material using a powder-bed deposition process. Courtesy of Blueprint Magazine UK.

Extrusion machines are programmed to deposit successive layers of a viscous mixture, typically raw earth or cement-based compounds, that solidify rapidly after deposition. One of the main constraints of this technique is gravity, which requires the printed forms to develop primarily along the vertical axis with little room for cantilevers. Unlike traditional wet techniques such as concrete casting, this method eliminates the need for formwork and reduces material curing times, streamlining on-site construction processes. Examples of architectural applications are *TECLA*, a 3D-printed house completed in 2021 and designed by Mario Cucinella Architects for WASP, which employed multiple printers operating simultaneously to extrude local soil (Moretti 2023); and *House Zero*,[11] completed in 2022 in Austin by Lake Flato Architects using ICON Build's

3D concrete printing technology, conceived as an on-site design-build proof-of-concept. Both projects reflect the convergence of technological innovation and architectural design quality, while also addressing challenges related to the performance of the building envelope.

AM technologies expand design possibilities by enabling the creation of complex morphologies. In this context, the digital 3D model is no longer just a representation but a source of knowledge about the building, directly informing production (Codarin 2023). These technologies are not yet affordable; they remain premium solutions. However, this is where research and development become active, with the ambition to digitize the construction site and prioritize design quality, while working toward making these technologies more accessible and contributing to a more technologically engaging valued role for construction labor.

Feedback-Loop Systems

Industrial automation often refers to machines capable of repeating programmed tasks. These tools are not necessarily aware of their surroundings and typically function as performative extensions of analog instruments. However, more advanced tools also exist, equipped with sensors or computer vision systems that can perceive and interpret their environment. Sensors, combined with computer vision systems, connect robots, humans, and materials through a nonlinear workflow driven by external information flows. This exchange of incoming and outgoing data creates a feedback loop informed by robotic sensing, which expands the robot's perceptual capabilities and responsiveness to its environment. Each time system conditions change, as often happens in dynamic environments like construction sites, the robot can adjust its movements to maintain a predetermined objective. In this way, the programming adapts in real time, prioritizing the optimization of outcomes despite shifting external factors. Human interaction with sensor-driven systems supports a hybrid approach that combines craftsperson intuition with digital analysis and robotic precision (Dubor et al. 2016). This dynamic is evident in AM, where feedback loops enable continuous adjustments across digital and physical parameters, such as material properties, extrusion flow rate, curing time, and overall performance (Sutjipto et al. 2019).

A robotic sensing workflow typically involves several interrelated components: sensors that collect real-world data, control protocols that adjust instructions in response to new inputs and manage the production sequence, online operations that transmit calibrated instructions to the robot's digital interface, and a compatible information platform that ensures continuous, coherent data flow throughout the process (Raspall et al. 2014). The feedback loop strategy in cyber-physical robotic systems (Menges 2015) makes it possible to manufacture complex material systems that are difficult to predict through digital simulation, due to the nonlinear relationship between material, manufacturing tool, and external parameters (Brugnaro et al. 2019). The inclusion of dimensional tolerances within the digital fabrication process reflects a convergence between the predictable digital world and the unpredictable physical reality. This type of nonlinear interaction and unpredictability often emerges in interventions on existing buildings as layers are uncovered and new conditions are revealed.

Cobots

Cobots, or collaborative robots, are tools designed to interact with humans within a shared workspace. They operate at low speeds and are equipped with sensors that allow them to detect obstacles, which enables safe collaboration. The term originates from the 1996 paper *Cobots: Robots for Collaboration with Human Operators*, by Northwestern University professors Edward Colgate and Michael Peshkin, who defined cobots as robotic devices intended to manipulate objects in coordination with a human operator (Colgate et al. 1996). Cobots were developed to advance the intelligent automation movement and overcome the idea of robots as obedient automated subordinates (Picon 2014). They operate autonomously and automatically, with the ability to be reprogrammed, typically using a fixed or mobile multi-purpose actuator functioning across three or more axes in industrial automation contexts. In the publication *Human +*

Machine: Reimagining Work in the Age of AI, Paul Daugherty and James Wilson emphasize that, unlike traditional robots, cobots are designed to work in close collaboration with humans. This enables manufacturers to rethink previously static processes, allowing workers to interact with intelligent machines and take on new, evolving roles (Wilson and Daugherty 2018). The integration of collaborative robotics aligns with the broader logic of digital transformation and provides a practical response to the ongoing decline in labor and skills within the construction sector, a trend that has persisted since the global market downturn of 2007. At the same time, such integration lays the groundwork for a reassessment of the industry through the lens of Industry 5.0 (European Commission 2021). Cobots transform the way humans and machines engage in manufacturing and, potentially, architectural production; an evolving relationship defined by continuous interaction and extended conversations among designers, builders, robots, and computational tools.

Augmented Reality in Building Construction

Augmented reality (AR), as a digital overlay onto the physical world, can serve as a bridge between computational operations and intuitive human input. This potential has been demonstrated through applications like those developed by Fologram, which created a system that guides users through complex task sequences using holographic instructions. An example is *Holographic Brickwork*, a parametrically generated wall constructed by bricklayers wearing AR glasses that projected real-time visualizations of each brick's position and sequence. The process transformed traditional labor into digitally assisted craftsmanship, allowing construction with precision directed by virtual feedback.

In 2019, the research group at Gramazio Kohler worked on a brick façade project for a winery in Greece that incorporated AR directly on-site. Workers placed bricks by following a custom dynamic optical guidance system with inertial object tracking (Mitterberger et al. 2020). The project echoed earlier work on the iconic *Winery Gantenbein*, which in 2011 pioneered full-scale digital robotic prefabrication. More recently, a

2023 example of AR-assisted construction used Microsoft HoloLens 2 to guide untrained workers in assembling a falsework-minimal masonry vault. Holographic overlays visualized each brick's placement in real time, eliminating the need for scaffolding or printed drawings (Oval et al. 2024).

As robots don't yet have the cognitive abilities to conduct tasks in full autonomy or contribute to decision-making, ongoing research is exploring how AR can act as a communication layer between computational models and robotic fabrication processes. Through real-time visualization, progress tracking, and interactive control, AR helps close the cognitive gap between digital design and physical execution (Makhataeva and Varol 2020; Suzuki et al. 2022). One example is the *2024 Cooperative Augmented Assembly* project, which introduced an experimental workflow combining human operators and robotic systems in custom timber construction. A HoloLens 2 application delivered step-by-step spatial instructions and sequential overlays, guiding users in accurately placing timber elements. This method challenged linear workflows by reintroducing human input during fabrication and allowing parameters to be adjusted on the fly. Humans performed tasks requiring dexterity and nuanced judgment, while robots followed up with precise, repeatable actions (Alexi et al. 2024).

AR-robotic integration creates space for keeping humans meaningfully involved in construction processes, bringing a more human-centered dimension to automation. Far from replacing entire processes or workflows, AR acts as a flexible enhancement, plugging into discrete tasks and augmenting them. These developments suggest a shift toward a more collaborative human-machine paradigm grounded in adaptive task sharing.

A Note on Artificial Intelligence

Since 2022, artificial intelligence has rapidly disrupted the fields of architecture and design, driven by the open access to image-generation platforms like Midjourney and Stable Diffusion, soon followed by advancements in video generation and 2D-to-3D translation technologies. AI has firmly entered the architectural discourse, prompting the urgent need for a new kind of visual literacy capable of engaging with these fast-evolving tools (Leach 2022). At present, these innovations are most impactful in the early stages of design, bringing immediate specificity and clarity to conceptual exploration (Bieg 2024; Koehler 2023). However, they fall short of fulfilling the promise of a seamless AI-to-construction pipeline, particularly in their integration with digital fabrication, which remains one of the most compelling prospects for the coming years.

Reflecting on how design and making have historically intersected, the history of architecture appears shaped by alternating moments of convergence and divergence. Similarly, early research in artificial intelligence and robotics emerged around the same time in the 1950s, initially without a clear distinction between the two disciplines. In the 1960s, many early cybernetic and AI projects set expectations beyond what the technology of the time could support, and by the early 1970s, the integration of AI and robotics appeared increasingly out of reach. Meanwhile, robotics began to align more closely to industrial automation, with focus on precision and repeatability instead of adaptability or intelligence (Chella et al. 2006). Fast forward to today, we are witnessing a renewed interest in the convergence of AI and robotics, an intersection largely set aside in past decades. Thanks to advances in computing power and data processing in the Fourth Industrial Revolution, many of the original ideas proposed by early AI research are finally becoming viable, bringing the conversation back to the forefront.

Despite their different origins, both AI and robotics can be seen as disembodied tools: the robot distances itself from its human operator through mechanical agency, while AI is detached from the physical world altogether. Yet not all AI systems require physical embodiment, and not all cognitive problems need to be extended into the material world. However, integrating AI and robotics, as a prototypical concept, forces AI to engage with real objects in real environments, turning it into an embodied agent (Brady 1985). In this scenario, robots are coupled with artificially intelligent algorithms that enhance their physical capacities while also expanding their cognitive abilities. Through real-time sensing and adaptive computational processes, these machines form an intelligent feedback loop that connects digital models, environmental inputs, and physical outputs in dynamic and responsive ways.

The implications of working with these advanced tools are discussed by Mario Carpo in a conversation with Jenny Sabin during the 2017 Fabricate conference, where he observes "artificial intelligence now suggests an almost pre-scientific intuitive approach to making. We really don't know much about this magic power of digital intuition. From a distance, it is clear that its closest parallel is the medieval master builder, the artisan who can manipulate materials and can conceive and make without designing" (Sheil et al. 2017). Paradoxically, the use of AI eliminates the need to experience phenomena directly, as computational power can simulate and predict outcomes in a machine-like manner, bypassing the need for individual empirical validation. This logic reflects a shift toward a post-humanist framework (Hayles 1999), wherein humans inform the machine and receive optimized, data-driven feedback in return.

AI robotics can be envisioned as intelligent agents capable of performing on-site operations during construction. Projects like *ImageNet* and *RoboNet*, originating from the Stanford Vision and AI Lab, provide a data infrastructure for such developments. These projects focus on creating vast image databases, categorized within hierarchical structures to facilitate machine learning and AI applications. Robots can be programmed with tailored data packages specific to the environment in which they operate. Informed by this data, robots can also interpret visual and spatial information, anticipate hazards, and contribute to design development through continuous feedback with architects, programmers, and builders. Comparable advances are also visible in other domains, for example, DeepMind's *Aeneas* project assists historians in reconstructing fragmentary Latin inscriptions by restoring missing text and contextualizing meaning, demonstrating how AI can extend disciplinary knowledge (Hall 2025). In both cultural heritage and construction, this approach points toward AI robotics as a paradigm of embodied intelligence, where a physical agent can reason about geometry, physics, and time-based sequences in order to operate autonomously within the built environment.

Robots: The Assembly Line Applied for Customized Architecture

The Origins

Many handbooks in the field, such as *Robot Technology Fundamentals* (Keramas 1998) or *The Robot: the Life Story of a Technology* (Nocks 2008), mention that the field of robotics has its origins in science fiction. The term "robot" comes from an English translation of the Czechoslovakian word "robota" which was coined by the writer Karel Capek (1890–1938) in his 1921 play *RUR—Rossum's Universal Robots*, where robots are depicted as humanoid slaves (Capek 2004). In Czech, robot means "boredom," "monotony," and "forced labor" alluding to repetitive tasks demanded by factory assembly lines (Carpo and Lemerle 2008). It was not until four decades after Čapek's play that the development of actual industrial robotics began. The term "robotics" itself was coined by Isaac Asimov in his 1942 science fiction story *Runaround*, which was first published in the journal *Astounding Science-Fiction*. In this story, Asimov formulated the now-famous Three Laws of Robotics: a robot may not harm a human being or, through inaction, allow a human to come to harm; a robot must obey orders given by humans unless such orders conflict with the First Law; and a robot must protect its own existence as long as this does not conflict with the First or Second Laws. Decades later, in his 1985 novel *Robots and Empire*, Asimov introduced the addition of a "Law Zero": a robot may not harm humanity, or, through inaction, allow humanity to come to harm. This broader ethical principle shifted the focus from individual humans to humanity as a whole.

Moving from sci-fi into reality, history reveals that the first industrial robot was tested within the automobile sector. It was based on the 1961 patent *Programmed Article Transfer* by American engineer George Devol, who introduced the concept of universal automation. This concept led to the creation of Unimation, the first robotics company, which he co-founded with Joseph Engelberger. The development of Devol's prototype culminated with "Unimate#001," a hydraulic manipulator arm that was installed in a General Motors factory in New Jersey to do welding and unload a die-castings press. Contrarily to Capek's humanoids and Asimov's artificial agents, the first automatic machine was neither a robot soldier nor a spacewalker. Unimate could repeat sequences of motions and was used to replace man in dangerous and extremely repetitive tasks described by the four Ds—dangerous, dull, dirty, and dumb—and four Hs—hot, heavy, hazardous, and humble (Keramas 1998). Compared to other manufacturing technologies, the robot was not considered a utensil since it was created to replace man. More advanced computer-controlled, sensor-driven electric arms were developed in the late 1960s and 1970s at both MIT and Stanford University. Since the 1990s, smaller electric arms, often referred to as service robots, have become increasingly common in fields such as medicine and advanced biology.

The Fourth Industrial Revolution continues to highlight the advancement of robots evolving alongside information technologies, with the electronics and automotive industries remaining the leading sectors (Fleck 2024). The *2024 World Robotics Report by International Federation of Robotics* shows an all-time high in the number of industrial robots automating production around the world and the annual number of units operating in factories worldwide increased by 10 percent in 2023.[12] The 1986 *Standard Handbook of Industrial Automation* reminds us that a widely accepted definition of robot is a "reprogrammable, multifunctional manipulator designed to move material, parts, tools, or specialized devices, through variable programmed motions for the performance of a variety of tasks" (Considine and Considine 1986). In other words, "the primary purpose of the robot as a machine is controlled motion: if it does not move, it is not a robot" (Keramas 1998). For the time being, the human mind puts them in motion.

Robots come in a wide variety of shapes, sizes, speeds, and load capacities. Their motion characteristics differ based on mechanical design. Commercial robots typically fall into five common categories of anatomy: rectangular or cartesian, cylindrical or post-type, spherical or polar, jointed-arm (also known as articulated or revolute), and SCARA, which stands

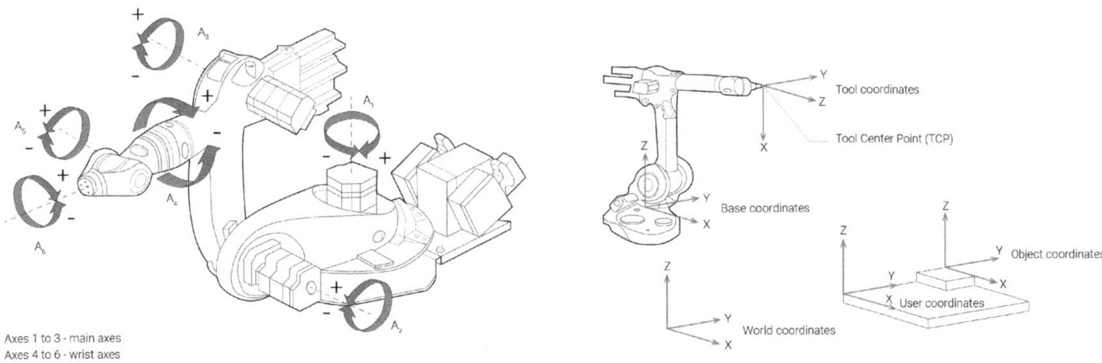

Figure 3.6 Geometry of a jointed-arm robot. Usually, based on the number of axes, joints are labeled: A1, A2, A3, A4, A5, and A6 allowing the robot to perform kinematics and interact with the surrounding three-dimensional environment. The tool center point (TCP) is the origin of the coordinate system or the point of action of the tool attached to the robotic arm.

for Selective Compliance Assembly Robot Arm. In industrial automation, jointed-arm robots are the most common type. These machines use rotary joints to connect rigid links, allowing them to reach any point in space in a way that mimics the movement of a human arm. This type of robot can move at high speeds in various directions and has a greater variety of angles of approach to a given point (Fig. 3.6). As James Keramas explains, "a robot can become a production machine only if a tool or device has been attached to its mechanical arm by means of the tool-mounting plate" (Keramas 1998). This attached device, known as an end-effector, defines the material processing task and transforms a generic robot into a task-specific machine. Depending on the type of operation, conventional end-effectors are equipped with various devices and tool attachments. Nevertheless, robots have no awareness or concern for the tasks they perform. This places responsibility on human professionals to use robotics thoughtfully, easing workers out of harmful or monotonous roles, and acknowledging that blind automation rarely leads to meaningful progress.

From Assembly Lines to Architecture

Robotics took time to find a meaningful place within architecture, and its integration into restoration practices is expected to be even more gradual. Identifying the absolute "first" architect to use robotics in architecture is difficult due to the field's evolving nature and collaborative aspects. However, leading university research labs have been central to adapting industrial robots for architectural applications, experimenting with new tools, workflows, and materials. In 2010, the Association for Robots in Architecture was founded with the aim of making robotic tools accessible to the creative industries. This has positioned the Association as a transdisciplinary network of robotic practitioners worldwide. Beginning in 2012, the Association launched the Rob|Arch conference series, Robotic Fabrication in Architecture, Art, and Design, which gathers international case studies and experimental practices, presenting robotic approaches across scales, techniques, and material systems.

Research in robotic fabrication and digital construction spans a wide range of strategies across major academic institutions. ETH Zurich's Chair of Architecture and Digital Fabrication, led by Fabio Gramazio and Matthias Kohler, has been at the forefront since the early 2000s. Their work integrates software, material research, and robotic programming into architectural workflows, tested through real-scale prototypes such as the *DFAB House*. Their publication *The Robotic Touch: How Robots Change Architecture* (Kohler et al. 2014) explores digital materiality, algorithmic thinking, and large-scale robotics, showing how architects have adapted manufacturing tools to redefine digital fabrication and human-machine collaboration. The Institute for Advanced Architecture of Catalonia emphasizes experimental construction methods using natural materials, advancing both digital tectonics and AM. At UCL Bartlett, research combines architectural theory with digital practice, focusing on computational workflows, automation, and fabrication strategies. At the University of Michigan, Wes McGee leads discourse on robotic fabrication, supporting the development of implementation workflows across various materials and experimental actuators. The Institute for Computational Design and Construction at the University of Stuttgart focuses on biomimetic and morphogenetic design, creating lightweight, adaptive structures inspired by natural systems, particularly through their ongoing series of research pavilions. The CREATE Lab at the University of Southern Denmark, led by Roberto Naboni, contributes to this evolving field by exploring AM with experimental materials and robotic fabrication of low-carbon, 3D-printed concrete structures. At the University of Pennsylvania, the Polyhedral Structures Laboratory, directed by Masoud Akbarzadeh, investigates robotic assembly and digital construction of discrete systems, bridging geometry, computation, and automation.

Robotics serves as a decisive expression of the CAD/CAM relationship, embodying the immediacy between conceptual thought and built reality (Picon 2014), forming a direct link between design and production in the architectural process. This connection is particularly relevant because, as Robert Aish explains in *First Build Your Tools*, "geometry and algorithms can exist in the abstract, but to be of any practical significance, to become a design tool which can be used by designers, then these have to be encapsulated in an executable form" (Aish et al. 2013).

Robotic Processes and Their Role in Higher Education

Culture of Technology at LTU CoAD

Lawrence Technological University, located in the Detroit metro area, was founded in the heart of the American industrial revolution. Its story begins in the 1930s, in the shadow of Henry Ford's Highland Park Plant, the birthplace of modern manufacturing and Fordism.[13] The first LTU students were non-traditional learners, workers from the factory floor who labored on assembly lines by day and filled classrooms by night. For them, education was not abstract, but a practical tool for entering the middle class. Building on this legacy, LTU embraces technology in the same regard—a valuable tool that upholds the integration of theory and practice, and acts as a defining ethos in the age of AI. At the College of Architecture and Design, a strong culture of making is embedded across disciplines. All students are required to learn the fundamentals of operating industrial robotic arms, and to develop the ability to "speak machine" (Maeda 2019), orchestrating workflows for intelligent fabrication. The college participates in the Fourth Industrial Revolution by using post-factory "retired" robots, machines that are dismissed from assembly lines as manufacturer production upgrades demand more advanced systems. The robots are repurposed to extend their lifespan in areas such as education, and their reduced cost reflects the growing body of technical and operational knowledge in automation.

Marty Neumeier, in *Meta Skills: Five Talents for the Robotic Age*, critiques the existence of a longstanding divide between thinking and making: "five-hundred years after the Renaissance, academic education in the West has been successful in separating the hand from the brain. We've decided that making things is less valuable than knowing things and therefore making has less place in the classroom. This is not only wrong, but it denies the very evolutionary advantage of being human." Neumeier argues that in today's world, making must be rejoined with knowing (Neumeier 2013). This reintegration is fundamental to architecture, a discipline historically defined by the fusion of theory and practice. Words that echo those of Vitruvius in the 1st century BCE, who defined practice as "the continuous and regular exercise of employment where manual work is done with any necessary material according to the design of a drawing," and theory as "the ability to demonstrate and explain the productions of dexterity on the principles of proportion" (Morgan and Warren 1914). In construction, craftspeople embody this synthesis, uniting theoretical knowledge with practical experience. Their learning is grounded in "doing," an approach described by Tim Ingold as "thinking through making," instead of simply applying preexisting theoretical concepts (Ingold 2013). An interpretation of learning by doing is inherent in the theorization of *Il Discorso Mentale*, where Leonardo Da Vinci speaks about the creative process to produce works of art such as paintings: "to Leonardo, an artist didn't learn to paint. He painted to learn" (Neumeier 2013).

Learning through Robotic Making

The College of Architecture and Design at LTU has developed multiple iterations for integrating robotics into the curriculum, aligning both theoretical frameworks and technical applications within a broader culture of making, where production is no longer treated as the final stage of design (Codarin and Daubmann 2021). Coursework is structured around exercises that support tool operation and the making of scale models. These forms are shaped by material constraints and surface behavior but guided by tactile knowledge (Ingold 2013). The approach responds to the lack of material engagement often found in BIM-based workflows and rapid prototyping, which tend to produce direct outputs with minimal physical interaction. In this context, the robot functions as a bridge between thinking, digital modeling, and the physical environment itself. As the workspace expands, so does the mental framework through which design is developed. Robots paired with design software support a two-way process between parts and models but rather than acting as passive machines, robots translate ideas into physical outcomes through immediate connections between three-dimensional digital inputs and real-world fabrication (Codarin 2024). Different end-effectors can be mounted on to the robot to perform varied tasks and allow for the same machine to take on multiple roles. Iterations can be produced with increasing complexity and refined fabrication strategies, further reinforcing the connection between thinking and making. Simulations within software environments such as Rhinoceros and Grasshopper help visualize motion paths and identify errors before moving into physical space.

The teaching structure introduces robotic fabrication through a progression of assignments, beginning with basic motion programming and advancing to material-based processes. Students simulate movement using 3D modeling software and generate toolpaths, working with coordinate systems and polar references. They also learn to jog the robot using a manual controller, teach pendant, which helps develop spatial awareness (Fig. 3.7). As skill building progresses, students begin to explore time as a design parameter through visual experiments like light painting with LEDs and long-exposure photography.[14] In this context, students incorporate AI large language models to script equations that drive robotic motion, shifting from geometry-based form-making to complex systems

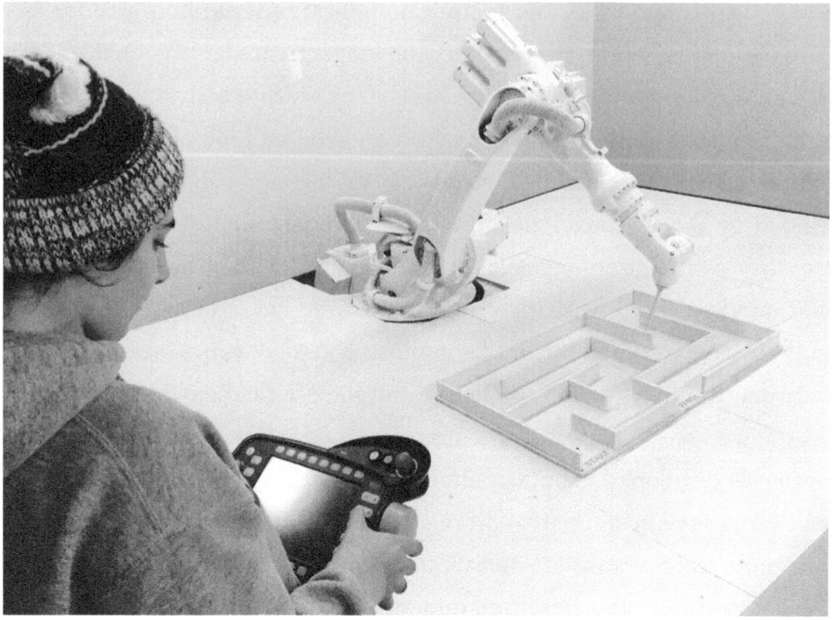

Figure 3.7 Robotic jogging exercise using a physical controller or teach pendant. The activity demonstrates the difference between jogging the robot in joint coordinates versus a world coordinate system. Photo by Sara Codarin.

shaped by mathematical logic (Fig. 3.8). Other assignments explore doubly curved surfaces and their relationship with normal vectors, building a deeper understanding of form through physical modeling (Fig. 3.9). More advanced exercises introduce clay printing (Fig. 3.10) and thermal cutting with a hot wire to produce simple proto-architectural forms, shaped through a sequence of calculated cuts on foam blocks (Fig. 3.11). These tasks allow students to iterate form generation while adjusting robotic parameters such as speed and temperature. Photographic documentation of the models supports further development through AI-driven image workflows, ranging from realistic rendering generation to retexturing surfaces, or creating new 3D forms by blending multiple 2D inputs (Fig. 3.12). Some projects extend into the fabrication of foam formwork for plaster casting, exploring the spatial relationship between void and solid and framing robotics as a dynamic design medium able to challenge conventional workflows with new modes of thinking and making.

In the design and construction, the role of the architect is shifting closer to that of the master builder, someone directly engaged

Figure 3.8 Robotic light painting captured through long-exposure photography. The geometry is derived from the Dequan Li attractor, a function from the mathematical field of dynamical systems. A large language model is used to generate Python code, which is then integrated into a Grasshopper script and exported via a robotic programming plugin, in this case Kuka|PRC. Student work by Nolan Barrios, Leo Georges, and Claudia Yakeera at LTU CoAD, spring semester 2023.

with construction and responsible for the full production chain of building elements. By working with robotics and digital fabrication tools, students gain insight regarding the distinction between upstream and downstream processes. Upstream processes can be defined as early stage design, where morphological decisions and strategic frameworks are established. Downstream processes follow and are tied to production methods, which in turn inform and enrich design through feedback and material constraints. A loose analogy can be drawn between robots and contractors: just as designers must clearly communicate intentions and constraints to a machine, they

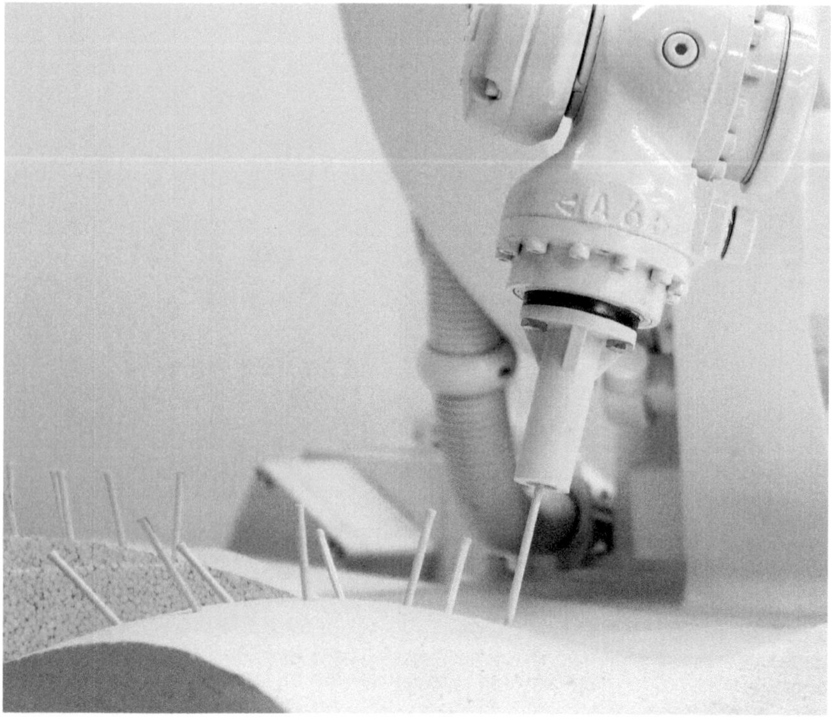

Figure 3.9 Robotic exercise involving the insertion of a grid of nails into a double-curved foam surface. The task is designed to explore robotic kinematics in relation to vectors guiding the motion in the three-dimensional space. Photo by Sara Codarin.

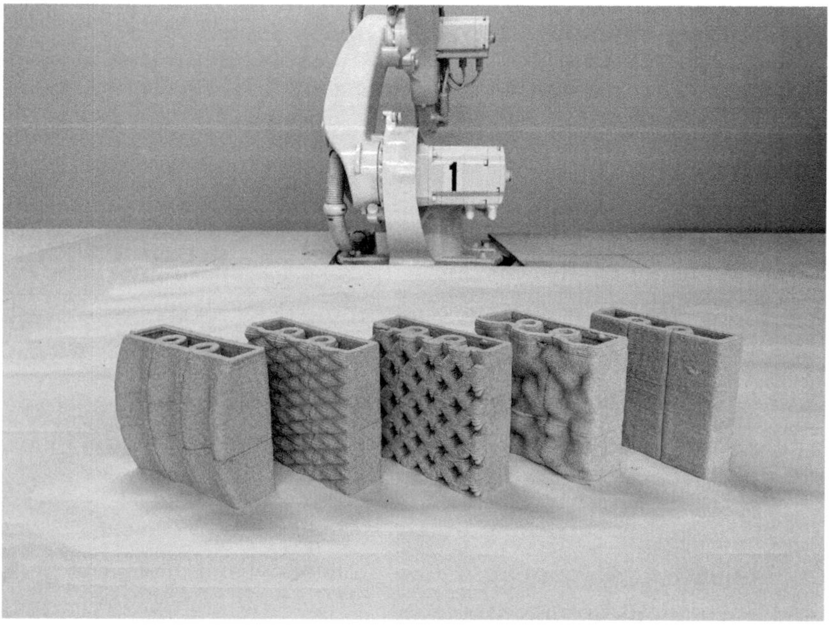

Figure 3.10 Robotically printed clay scale models of parametrically designed wall samples. Photo by Sara Codarin.

Figure 3.11 Robotic hot wire fabrication process used to create a series of massing models or proto-architectural forms. Student work from the spring semester of 2019. Photographic documentation is used as input for generating realistic renderings.

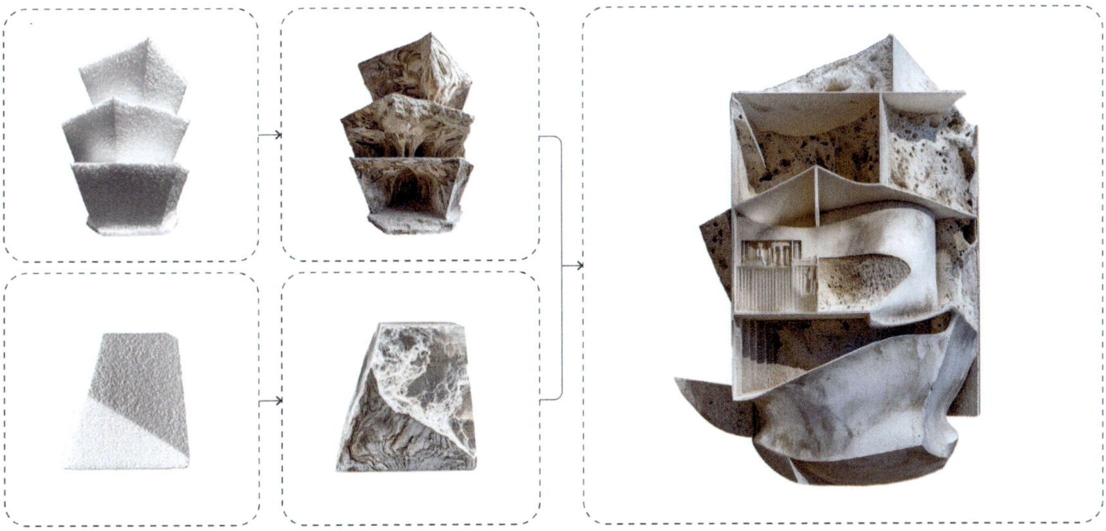

Figure 3.12 Robotically hot-wire-cut massing models. Student work from the spring semester of 2022. The physical forms are combined three-dimensionally and retextured to produce a speculative artifact that explores materiality following the prototyping process.

must also be able to articulate design goals and fabrication logic to contractors or manufacturers. Through digital tools, architects can develop a stronger understanding of production systems, allowing technical complexities to be addressed earlier in the design process, leading to more informed decisions. In this context, students learn to engage with prototyping as a convergence of design and production. The workspace itself becomes hybrid, both physical and digital, what might be called "phygital" (Yan et al. 2024). This continuous exchange between digital models and material output becomes central to the pedagogy that students are encouraged to develop, reflecting on formal ideas shaped by the tools and workflows that support them. Mediating between design and material becomes a key skill, allowing speculative models to emerge from digital inputs while also reflecting a new relationship between future designers and the technologies they use.

Notes

1 Karl Daubmann is a Fellow of the American Institute of Architects (FAIA) and was the 2017 Architecture Fellow at the American Academy in Rome (FAAR).

2 Source: Macrina Wilkins, Senior Research Analyst, AGC of America, from Bureau Economic Analysis (GDC); Census Bureau (spending); Bureau of Labor Statistics (national and state employment), (median wages). September 27, 2024.

3 EU-robotics roadmap: https://eu-robotics.net/topic-groups-and-the-roadmap-2030.

4 The concept of digital complexity was explored by Benjamin Dillenburger with the project "Digital Grotesque". Another example is the work "Quaquaversal Centrepiece at the Spring–Summer 2016 Iris Van Herpen ready-to-wear-collection" where robotic arms 3D printed and manipulated a dress. The project was carried out by Iris van Harpen, Jolan van der Wiel, and Marjian Colletti + Rexilab.

5 The only existing Dymaxion House prototype is on display at the Ford Museum of American Innovation in Dearborn, Michigan. See: https://www.thehenryford.org/visit/henry-ford-museum/exhibits/dymaxion-house/.

6 An example of single task robot is the Z-Hand vertical delivery system for transporting material for interior finishing. The system was developed by the company Kajima.

7 An example of integrated automation is the SMART system developed by Shimizu Corporation, which combines pick-up, vertical transport, and precise positioning into a unified process.

8 An example of a sky factory system is Akatuki 21, developed by Fujita for the construction of steel-based buildings. The system integrates material delivery, on-site logistics, and the assembly of structural components.
9 "Designing Out Waste. A Brief on Robotic Refabrication" is the culmination of the inaugural FPInnovations Scholar-in-Residence program at the School of Architecture Planning + Landscape (SAPL) in collaboration with founder of GXN Kasper Guldager Jensen as visiting professor at the University of Calgary.
10 Stereolithography (SLA) is a 3D printing technique that uses a laser to solidify liquid photopolymer resin into plastic, building the object one layer at a time.
11 House Zero in Austin, Texas, is a showcase project by ICON that highlights the use of 3D printing in home construction. To test a more scalable model, ICON also built Wolf Ranch, a community of 100 3D-printed homes designed by Bjarke Ingels Group. It's the largest project of its kind so far.
12 Source: International Federation of Robotics (IFR), World Robotics 2024 Report. Accessed from: https://ifr.org/ifr-press-releases/news/record-of-4-million-robots-working-in-factories-worldwide.
13 Fordism refers to a system of mass production and consumption that shaped much of the mid-20th century. It relied on assembly line manufacturing, standardization, and specialization to enable large-scale production and broad consumer access. Pioneered by Henry Ford, the model aimed to increase efficiency and reduce costs, making goods more affordable.
14 These experiments draw from the time and motion study of Frank and Lillian Gilbreth in the early 20th century, who used visual recording to analyze and improve labor efficiency.

References

Aish, Robert, Brady Peters, and Terri Peters. 2013. "First Build Your Tools." In *Inside Smartgeometry: Expanding the Architectural Possibilities of Computational Design*, 1st ed. Wiley. https://doi.org/10.1002/9781118653074.

Alexi, Eleni Vasiliki, Joseph Clair Kenny, Lidia Atanasova, Gonzalo Casas, Kathrin Dörfler, and Daniela Mitterberger. 2024. "Cooperative Augmented Assembly (CAA): Augmented Reality for on-Site Cooperative Robotic Fabrication." *Construction Robotics* 8 (2): 28. https://doi.org/10.1007/s41693-024-00138-6.

Arbizzani. 2015. *Tecnica e Tecnologia dei Sistemi Edilizi: Progetto e Costruzione*. Maggioli.

Baker, Kermit, and Jennifer Riskus. 2008. *AIA Compensation Report 08: A Survey of U.S. Architecture Firms*. American Institute of Architects.

Bieg, Kory. 2024. "The Etiology of a New Collective Architecture." In *Diffusions in Architecture*, Edited by Matias Del Campo, 1st ed. Wiley. https://doi.org/10.1002/9781394191802.ch8.

Bock, Thomas, and Silke Langenberg. 2014. "Changing Building Sites: Industrialisation and Automation of the Building Process." *Architectural Design* 84 (3): 88–99. https://doi.org/10.1002/ad.1762.

Bock, Thomas, and Thomas Linner. 2016. *Site Automation: Automated/Robotic On-site Factories*. Cambridge University Press.

Brady, Michael. 1985. "Artificial Intelligence and Robotics." *Artificial Intelligence* 26 (1): 79–121.

Brell-Çokcan, Sigrid, and Johannes Braumann, eds. 2013. *Rob/Arch 2012: Robotic Fabrication in Architecture, Art and Design*. Rob/Arch 2012, Wien; New York. Springer.

Brugnaro, Giulio, Angelo Figliola, and Alexandre Dubor. 2019. "Negotiated Materialization: Design Approaches Integrating Wood Heterogeneity Through Advanced Robotic Fabrication." In *Digital Wood Design*, Edited by Fabio Bianconi and Marco Filippucci, vol. 24. Lecture Notes in Civil Engineering. Springer International Publishing. https://doi.org/10.1007/978-3-030-03676-8_4.

Capek, Karel. 2004. *RUR (Rossum's Universal Robots)*. Penguin.

Carpo, Mario, and Frédérique Lemerle. 2008. *Perspective, Projections and Design: Technologies of Architectural Representation*. Routledge.

Chella, Antonio, Luca Iocchi, Irene Macaluso, and Daniele Nardi. 2006. "Artificial Intelligence and Robotics." *Intelligenza Artificiale* 3 (1–2): 87–93.

Cicconi, Ivan. 1981. "La Struttura Produttiva: Stato Attuale e Possibili Evoluzioni." In *Progettare nel Processo Edilizio. La Realta Come Scenario per l'Edilizia Residenziale*, Edited by Mario Zaffagnini. Luigi Parma.

Codarin, Sara. 2023. "Enhancing the Workforce in Construction: Robotic Concrete Printing in Detroit." *TECHNE – Journal of Technology for Architecture and Environment*, no. 25 (May): 233–42. https://doi.org/10.36253/techne-13704.

Codarin, Sara. 2024. "Prototyping and Fabrication in Design Education." *Proceedings of the National Conference on the Beginning Design Student (NCBDS): Shared Resources* (Lafayette, Louisiana), February 29.

Codarin, Sara, and Karl Daubmann. 2021. "Speculative Physical Models Created Through a Robotic Process." *Proceedings of Cumulus: Design Culture(s)* (Rome, Italy), June 8, 461–75.

Colgate, J. Edward, Witaya Wannasuphoprasit, and Michael A. Peshkin. 1996. "Cobots: Robots for Collaboration with Human Operators." *Dynamic Systems and Control*, November 17, 433–39. https://doi.org/10.1115/IMECE1996-0367.

Considine, Douglas M., and Glenn D. Considine. 1986. "Robot Technology Fundamentals." In *Standard Handbook of Industrial Automation*, Edited by Douglas M. Considine and Glenn D. Considine. Springer US. https://doi.org/10.1007/978-1-4613-1963-4_17.

De Kestelier, Xavier. 2011. "Design Potential for Large-Scale Additive Fabrication." In *Fabricate Making Digital Architecture*, Edited by R. Glynn and B. Sheil. UCL Press.

De Kestelier, Xavier, and Robert Buswell. 2009. "A Digital Design Environment for Large-Scale Rapid Manufacturing." *Proceedings of the 29th Annual Conference of the Association for Computer Aided Design in Architecture (ACADIA): reForm.* Chicago, Illinois, USA, 201–8.

Dörfler, Kathrin, Norman Hack, Timothy Sandy, et al. 2019. "Mobile Robotic Fabrication beyond Factory Conditions: Case Study Mesh Mould Wall of the DFAB HOUSE." *Construction Robotics* 3 (1–4): 53–67. https://doi.org/10.1007/s41693-019-00020-w.

Dubor, Alexandre, Guillem Camprodom, Gabriel Bello Diaz, et al. 2016. "Sensors and Workflow Evolutions: Developing a Framework for Instant Robotic Toolpath Revision." In *Robotic Fabrication in Architecture, Art and Design 2016*, Edited by Dagmar Reinhardt, Rob Saunders, and Jane Burry. Springer International Publishing. https://doi.org/10.1007/978-3-319-26378-6_33.

Dubor, Alexandre, Jean-Baptiste Izard, Edouard Cabay, Aldo Sollazzo, Areti Markopoulou, and Mariola Rodriguez. 2019. "On-Site Robotics for Sustainable Construction." In *Robotic Fabrication in Architecture, Art and Design 2018*, Edited by Jan Willmann, Philippe Block, Marco Hutter, Kendra Byrne, and Tim Schork. Springer International Publishing. https://doi.org/10.1007/978-3-319-92294-2_30.

European Commission. 2021. *Industry 5.0: Towards a Sustainable, Human Centric and Resilient European Industry.* Directorate General for Research and Innovation. Publications Office. https://data.europa.eu/doi/10.2777/308407.

Fleck, Anna. 2024. "Robotics and Automization: The Giants of Industrial Robotics." *Statista*, May 13.

Hall, Peter. 2025. "Google DeepMind's New AI Can Help Historians Understand Ancient Latin Inscriptions." *MIT Technology Review*, July 23. https://www.technologyreview.com/2025/07/23/1120574/deepmind-ai-aeneas-helps-historians-interpret-latin-inscriptions/.

Hayles, N. Katherine. 1999. *How We Became Posthuman: Virtual Bodies in Cybernetics, Literature, and Informatics*. University of Chicago Press.

Ingold, Tim. 2013. *Making: Anthropology, Archaeology, Art and Architecture*. Routledge.

Jensen, Kasper Guldager, Josh Taron, et al. 2019. *Designing Out Waste*. School of Architecture, Planning and Landscape, University of Calgary.

Keating, Steven, Nathan A. Spielberg, John Klein, and Neri Oxman. 2014. "A Compound Arm Approach to Digital Construction: A Mobile Large-Scale Platform for On-Site Sensing, Design, and Digital Fabrication." In *Robotic Fabrication in Architecture, Art and Design 2014*, Edited by Wes McGee and Monica Ponce De Leon. Springer International Publishing. https://doi.org/10.1007/978-3-319-04663-1_7.

Keramas, James G. 1998. *Robot Technology Fundamentals*. Delmar Publishers.

Khoshnevis, Behrokh. 2004. "Automated Construction by Contour Crafting—Related Robotics and Information Technologies." *Automation in Construction* 13 (1): 5–19. https://doi.org/10.1016/j.autcon.2003.08.012.

Koehler, Daniel. 2023. "More than Anything: Advocating for Synthetic Architectures within Large-Scale Language-Image Models." *International Journal of Architectural Computing* 21 (2): 242–55. https://doi.org/10.1177/14780771231170455.

Kohler, Matthias, Fabio Gramazio, Jan Willmann, et al., eds. 2014. *The Robotic Touch: How Robots Change Architecture; Gramazio & Kohler Research ETH Zurich 2005-2013*. Park Books.

Labonnote, Nathalie, Anders Rønnquist, Bendik Manum, and Petra Rüther. 2016. "Additive Construction: State-of-the-Art, Challenges and Opportunities." *Automation in Construction* 72 (December): 347–66. https://doi.org/10.1016/j.autcon.2016.08.026.

Leach, Neil. 2022. *Architecture in the Age of Artificial Intelligence: An Introduction to AI for Architects*. Bloomsbury Visual Arts.
Maeda, John. 2019. *How to Speak Machine: Computational Thinking for the Rest of Us*. International Edition. Portfolio//Penguin.
Makhataeva, Zhanat, and Huseyin Varol. 2020. "Augmented Reality for Robotics: A Review." *Robotics* 9 (2): 21. https://doi.org/10.3390/robotics9020021.
McGee, Wes, and Monica Ponce De Leon, eds. 2014. *Robotic Fabrication in Architecture, Art and Design 2014*. Springer International Publishing. https://doi.org/10.1007/978-3-319-04663-1.
Menges, Achim. 2012. "Biomimetic Design Processes in Architecture: Morphogenetic and Evolutionary Computational Design." *Bioinspiration & Biomimetics* 7 (1): 015003. https://doi.org/10.1088/1748-3182/7/1/015003.
Menges, Achim. 2015. "The New Cyber-Physical Making in Architecture: Computational Construction." *Architectural Design* 85 (5): 28–33. https://doi.org/10.1002/ad.1950.
Mitterberger, Daniela, Kathrin Dörfler, Timothy Sandy, et al. 2020. *Augmented Bricklaying: Human–Machine Interaction for in Situ Assembly of Complex Brickwork Using Object-Aware Augmented Reality*. Construction Robotics 4, no. 3 (2020): 151–161. https://doi.org/10.3929/ETHZ-B-000466341.
Moretti, Massimo. 2023. "WASP in the Edge of 3D Printing." In *3D Printing for Construction with Alternative Materials*, Bárbara Rangel, Ana Sofia Guimarães, Jorge Lino, and Leonardo Santana. Springer.
Morgan, Morris Hicky, and Herbert Langford Warren. 1914. *Vitruvius: The Ten Books on Architecture*. Harvard University Press.
Morgante, A. 2011. "Radiolaria Pavilion." In *Fabricate Making Digital Architecture*, Edited by R. Glynn and B. Sheil. UCL Press.
Naboni, Roberto, and Ingrid Paoletti. 2015. *Advanced Customization in Architectural Design and Construction*. Springer.
Nardi, Guido. 1980. *Tecnologia dell'Architettura e Industrializzazione nell'Edilizia*. 2nd ed. Ricerche di Tecnologia dell'Architettura 11. Franco Angeli.
NCARB. 2024. *NCARB By the Numbers – 2024 Edition*. https://www.ncarb.org/sites/default/files/NBTN-2024.pdf.
Neumeier, Marty. 2013. *Metaskills: Five Talents for the Robotic Age*. New Riders.
Nocks, Lisa. 2008. *The Robot: The Life Story of a Technology*. Johns Hopkins University Press.
Oval, Robin, Rafael Pastrana, Edvard P.G. Bruun, et al. 2024. "An Integrated Digital Design and Construction Approach for Falsework-Minimal Masonry Vaults." *Structures* 63 (May): 106428. https://doi.org/10.1016/j.istruc.2024.106428.
Petersen, Kirstin H., Nils Napp, Robert Stuart-Smith, Daniela Rus, and Mirko Kovac. 2019. "A Review of Collective Robotic Construction." *Science Robotics* 4 (28): eaau8479. https://doi.org/10.1126/scirobotics.aau8479.
Picon, Antoine. 2014. "Robots and Architecture: Experiments, Fiction, Epistemology." *Architectural Design* 84 (3): 54–59. https://doi.org/10.1002/ad.1754.
Raspall, Felix, Felix Amtsberg, and Stefan Peters. 2014. "Material Feedback in Robotic Production: Plastic and Elastic Formation of Materials for Reusable Mold-Making." In *Robotic Fabrication in Architecture, Art and Design 2014*,

Edited by Wes McGee and Monica Ponce De Leon. Springer International Publishing. https://doi.org/10.1007/978-3-319-04663-1_23.

Reinhardt, Dagmar, Rob Saunders, and Jane Burry, eds. 2016. *Robotic Fabrication in Architecture, Art and Design 2016*. Springer International Publishing. https://doi.org/10.1007/978-3-319-26378-6.

Sheil, R., Achim Menges, Ruairi Glynn, and Marilena Skavara. 2017. *Fabricate 2017*. UCL Press.

Sousa, José Pedro, Cristina Gassó Palop, Eduardo Moreira, et al. 2016. "The SPIDERobot: A Cable-Robot System for On-Site Construction in Architecture." In *Robotic Fabrication in Architecture, Art and Design 2016*, Edited by Dagmar Reinhardt, Rob Saunders, and Jane Burry. Springer International Publishing. https://doi.org/10.1007/978-3-319-26378-6_17.

Spadolini, Pier Luigi. 1981. *Progettare nel Processo Edilizio: la Realtà Come Scenario per l'Edilizia Residenziale*. Edited by Mario Zaffagnini. Luigi Parma.

Sutjipto, Sheila, Daniel Tish, Gavin Paul, Teresa Vidal-Calleja, and Tim Schork. 2019. "Towards Visual Feedback Loops for Robot-Controlled Additive Manufacturing." In *Robotic Fabrication in Architecture, Art and Design 2018*, Edited by Jan Willmann, Philippe Block, Marco Hutter, Kendra Byrne, and Tim Schork. Springer International Publishing. https://doi.org/10.1007/978-3-319-92294-2_7.

Suzuki, Ryo, Adnan Karim, Tian Xia, Hooman Hedayati, and Nicolai Marquardt. 2022. "Augmented Reality and Robotics: A Survey and Taxonomy for AR-Enhanced Human-Robot Interaction and Robotic Interfaces." *CHI Conference on Human Factors in Computing Systems*, April 29, 1–33. https://doi.org/10.1145/3491102.3517719.

Willmann, Jan, Federico Augugliaro, Thomas Cadalbert, Raffaello D'Andrea, Fabio Gramazio, and Matthias Kohler. 2012. "Aerial Robotic Construction towards a New Field of Architectural Research." *International Journal of Architectural Computing* 10 (3): 439–59. https://doi.org/10.1260/1478-0771.10.3.439.

Willmann, Jan, Philippe Block, Marco Hutter, Kendra Byrne, and Tim Schork, eds. 2019. *Robotic Fabrication in Architecture, Art and Design 2018: Foreword by Sigrid Brell-Çokcan and Johannes Braumann, Association for Robots in Architecture*. Springer International Publishing. https://doi.org/10.1007/978-3-319-92294-2.

Wilson, H. James, and Paul R. Daugherty. 2018. *Human + Machine: Reimagining Work in the Age of AI*. Harvard Business Review Press.

Yan, Chao, Hua Chai, Tongyue Sun, and Philip F. Yuan, eds. 2024. *Phygital Intelligence: Proceedings of the 5th International Conference on Computational Design and Robotic Fabrication (CDRF 2023)*. Computational Design and Robotic Fabrication. Springer Nature.

Zhang, Ketao, Pisak Chermprayong, Feng Xiao, et al. 2022. "Aerial Additive Manufacturing with Multiple Autonomous Robots." *Nature* 609 (7928): 709–17.

Chapter 4

New Technologies for the Conservation of Cultural Heritage

DOI: 10.4324/9781003590873-4

Traditional Approach to Architectural Heritage

Restoration is the practice of preserving objects of recognized value from decay, extending their lifespan without erasing the passage of time (Carbonara 2012, 1997). Unlike renovation, repair, or adaptive reuse driven by practical concerns, restoration focuses on the inherent value of historic artifacts, which are regarded as "documents," unique expressions of material culture worth passing on to future generations. In Italy, the theory of restoration has historically held a prominent position, shaped by the country's deep-rooted architectural heritage and the layered history of its cities. Some early restoration principles were codified during the Renaissance, particularly through the emergence of a historical-philological approach to studying the past (Lorusso 2002). Theoretical developments in Italy were also influenced in various ways by broader European discourse over time. In the 18th century, Johann Winckelmann's (1717–1768) contributions to archaeological science helped define a modern understanding of monument restoration. In the 19th century, Viollet-le-Duc (1814–1879) promoted the idea of stylistic restoration, which disregards the marks of time and instead seeks to recover the formal qualities of the monument's original era and place (Dalla Costa 2005). In contrast, John Ruskin (1819–1900) rejected restoration altogether, describing it as a destructive act rather than a means of preservation. For Ruskin, any intervention constituted tampering; he instead advocated for conservation in its strictest form, supporting minimal, consolidating measures intended only to slow the process of deterioration (Chatterjee 2019).

The guiding principles of restoration have developed through extensive theoretical debate, resulting in a set of key criteria that inform best practices. One such criterion is the distinguishability of intervention, which ensures that any additions to the original artifact remain clearly recognizable, avoiding the distortion of its historical interpretation. Reversibility is a closely related principle, derived from the Latin *revertere*, which implies that any intervention should allow for a return to the artifact's prior state, thereby permitting future updates or revisions. In this context, every restoration hypothesis must be both verifiable and open to amendment.

Authenticity is equally central, with restoration efforts rejecting any form of falsification. The principle of minimum intervention further affirms the need to act only when necessary for conservation, aiming to alter the existing structure as little as possible. Finally, any introduced materials must be physically and chemically compatible with the original fabric of the building to maintain long-term stability and integrity.

In the article, originally written in Italian, *Architectural Restoration: Evolving Concepts Through the Centuries*, key theoretical contributions are outlined that have shaped current restoration principles and discourse (Lorusso et al. 2002). Among them, Camillo Boito (1836–1914) is recognized as a foundational voice in the development of modern restoration. In response to the prevailing approaches of his time—stylistic restoration (return a building to its original style), as advocated by Viollet-le-Duc and conservative restoration (preserve a building as much as possible without alterations), promoted by John Ruskin, led to Camillo Boito proposing a "third way" (*terza via*), later known as scientific restoration (*restauro scientifico*). Diverging from both stylistic idealization and absolute conservation, Boito's approach is grounded in the careful study of the artifact and the historical sequence of its transformations. This analysis serves as the basis to determine the appropriate intervention, which may include individual or combined approaches such as consolidation to ensure structural stability; reconstruction or anastylosis to reassemble original elements to recover their historical form; removal of components that compromise structural or aesthetic integrity; or the integration of missing parts.

Boito's positions were further developed by Gustavo Giovannoni (1876–1946), who expanded the idea of scientific restoration and introduced the concept of philological restoration. In this approach, restoration should preserve the building's functionality, and when additions are required, they should replicate the missing forms as faithfully as possible while employing a simplified architectural language to ensure their distinguishability. Restoration theory further evolved through the ideas of Cesare Brandi (1906–1988) and his concept of critical restoration (*restauro critico*), articulated in his *Theory of Restoration*. In particular, Brandi called attention to the need to balance historical and aesthetic considerations in the restoration of cultural heritage. His notion of the "work's potential oneness" (*unità potenziale dell'opera*) suggests that restoration should aim to recover the original unity of the artifact, but only insofar as it avoids falsification or the erasure of the traces of time (Brandi 1963).

Most recently, the ideas of restoration have been articulated by Giovanni Carbonara (1942–2023) through his formulation of critical-conservative restoration. Critical, for the conviction that every intervention constitutes an "episode" in itself, not to be classifiable within fixed categories. Conservative, because it requires the architectural object to be passed on to future stewards. The widely accepted view is that buildings of historic value are high-priority elements within the built environment and must be analyzed on a case-by-case (*caso per caso*) basis. In this way, each architectural work must be interpreted within its historical and cultural context, recognizing that restoration is a direct material act through which preservation and conservation are implemented—potentially including modifications, provided they are guided by rigorous scientific and critical oversight.

Influential Charters within Conservation and Restoration Practices

An international effort to institutionalize the principles of conservation and the restoration of architectural heritage was the *1931 Athens Charter*, a manifesto that emerged from the First International Congress of Architects and Technicians of Historic Monuments and was inspired by the theories of Boito and Giovannoni. The charter called for national legislation to protect historic sites and it advanced the idea of a shared global heritage, reaffirming that monuments must be understood within their local histories to serve the broader human community, while still encouraging the careful and "judicious" integration of modern materials into conservation efforts.

The *Athens Charter* was followed by the *1932 Italian Restoration Charter* that supported a principle where, in the case of ancient monuments, any form of completion should generally be excluded. The *Italian Charter* endorsed the principles expressed in Athens and advocated for the use of modern technology to support scientific restoration, provided it aligns with a deep respect for the historical and aesthetic character of the work. The charter recognized that the most advanced construction techniques (*mezzi costruttivi modernissimi*) may offer valuable assistance and should be employed when traditional methods, those "analogous to the ancient ones," fail to achieve intended goals.

The *Venice Charter of 1964*[1] remains the most influential and comprehensive international document for conservation. This document codifies widely accepted standards for the preservation of architectural heritage and historic sites, establishing principles grounded in the concept of authenticity and promoting a unified approach across different countries and cultures. Building on the foundations of the *Italian Restoration Charter*, the *Venice Charter* clearly states that the process of restoration must stop at the point where conjecture begins (Art. 9). In cases where traditional techniques are insufficient, the consolidation of a monument may be carried out using modern methods, provided their effectiveness is supported by scientific evidence and proven through experience.

Approaches to the modernization of reconstruction processes in cultural heritage were also addressed in the *Italian Restoration Charter of 1972*, which adopted a top-down model, having been issued by the Ministry of Public Education. While the charter permitted the use of new procedures and materials in restoration, any such intervention required approval from the Ministry. The approvals were based on the recommendation of the Central Institute for Restoration (*Istituto Centrale del Restauro*), founded by historians Cesare Brandi and Carlo Argan. The institute was also responsible for advising on materials and techniques, proposing innovative procedures, and defining research initiatives.

Subsequent documents built upon these principles, such as the *Declaration of Amsterdam of 1975* that emphasized new materials and techniques to be tested and approved by independent scientific institutions before use and the *Lausanne Charter of 1990*, which encouraged the conservation of archaeological sites to lean on academic research and education as key tools for deepening disciplinary knowledge. Regarding integration of data collection, the *Declaration of Cooperation on Advancing Digitization of Cultural Heritage of 2019*, signed by the Member States of the European Union, identified the conservation and digital documentation of heritage as essential for its transmission to future generations.

Recurring themes in restoration discourse[2] over the decades include attention to both technological developments and aesthetic languages applied to interventions. These are often framed in broad terms, allowing space for critical interpretation. Early 20th-century charters established foundational restoration principles and emphasized a conservative approach. This is particularly evident given that many of the drafting committee members were European, and their perspectives were conditioned by the widespread destruction across the continent during the world wars. An example is Warsaw, Poland, which was almost entirely destroyed during World War II and later rebuilt *com'era, dov'era*, "as it was, where it was," as a means of mitigating the social trauma of territorial loss. A similar philosophy guided the reconstruction of the La Fenice Theatre in Venice, which was devastated by fire in 1996 and faithfully rebuilt under the supervision of Aldo Rossi. More broadly, cultural heritage across the globe will continue to face traumatic losses due to war, environmental disasters, and the accelerating effects of climate change. While what is lost cannot be recovered, these moments create opportunities to advance knowledge around reconstruction. As the construction industry and material sciences continue to evolve, our approaches to rebuilding must also adapt. Aesthetically, restoration has often favored simplified architectural languages that allow for recognizability. One codified technique is *tratteggio*, used in fresco restoration, where missing areas are filled in with fine, parallel brushstrokes to distinguish the restorer's hand from the original artist's. In masonry, the *cuci-scuci* method, literally "sew and unsew," consists of removing damaged sections of a wall (*scuci*) and reconstructing them with historically compatible materials (*cuci*) to restore both structural integrity and visual continuity. When it comes to surfaces, particularly within the framework of the potential unity of a work of art or architecture, patina is regarded as evidence of the passage of time. Preserving it is essential, as the more historical material is retained, the more the intervention respects the architectural artifact. These techniques have become standard practice. The question now is how digital design and fabrication technologies might enter this process without disrupting the values that underpin restoration work.

Innovating Technology and Managing Tradition

The contrast between digital design technologies and traditional material craft techniques used to construct historic buildings reveals a productive type of tension. Traditional methods and new processes represent two valid design trajectories: they can remain distinct or be thoughtfully hybridized, as long as they adhere to core principles of restoration. When approached in this way, combining past and present techniques enables a meaningful exchange with history, allowing contemporary interventions to engage in dialogue with existing works of architecture. The guiding principle in this process is the case-by-case logic of evaluation, which requires specificity to the context of each project and reconciles the use of advanced technologies with the integration of new technological materials. This approach relies on accumulated knowledge, design experience, and the architect's critical judgment to drive major design decisions on the construction site (Sposito 2004). Through a value-based lens, the architect navigates transformation by weighing historical significance alongside the potential for innovation (Tafuri 1991) and aims to align the value of heritage with the value of the new (Dal Co 2013).

Maria Luisa Germanà, in her paper *Technology and Architectural Heritage: Dynamic Connections*, argues that the innovations and unprecedented advances of scientific knowledge brought about by the Fourth Industrial Revolution have opened new possibilities for applying technological approaches to the conservation of architectural heritage and archaeological sites (Germanà 2019). In framing what she calls the "imperative of responsibility" in conservation, Germanà reflects on how the Third Industrial Revolution introduced quality-oriented tools for lean production, and she draws a parallel to how current technological shifts are gradually transforming both the cultural and operational dimensions of heritage work. Within this evolving context, she identifies "reliability" as a key guiding principle. This supports a legitimate ambition to connect digital fabrication, construction automation, and restoration—responding through nonstandard production methods to the complex needs of recovery sites where the rigid linearity of traditional building phases may fall short.

The ubiquity of digital infrastructure offers an opportunity to overcome rigid procedural constraints while still maintaining ties to traditional methodologies. For example, chisels and robots don't exclude each. Mobile robots and drones can be deployed on construction sites for data collection, and site-printing robots offer targeted, single-task automation. While fully automated on-site implementation is still developing, off-site robotic fabrication in smart facilities is already well established. Some restoration projects are leveraging this capacity, particularly in the reproduction of columns, sculptures, and ornamental features. Survey tools such as laser scanners and LiDAR sensors contribute both geometric and diagnostic data through non-invasive methods, enhancing knowledge documentation of buildings. However, this degree of digitization does not yet provide a fully practical decision-making toolkit or resolve complex, context-specific critical conditions (Germanà 2014), such as material decay patterns, structural instability, past undocumented interventions, or the ethical implications of reconstructing missing elements. Many interventions that demand high precision still depend on manual craftsmanship, reinforcing the reality that restoration work remains heavily reliant on skilled labor.

Assuming that restoration sites are rich with intelligence, craftsmanship, and tacit knowledge, the question becomes how to enable a smart reconfiguration of these environments, where tools and information are interconnected, and decisions are informed by translated digital data. In this scenario, industrial robots offer operational flexibility and the precision of disembodied tools that can be cognitively linked to data from digital surveys. Adjustments to programmed interventions can be made in real time through feedback loops, merging the physical and the digital and defining the operational framework of the "phygital heritage" (Nofal et al. 2017). These variables point to the need for well-defined workflows, where recursive loops allow for continuous adjustment in a solution-oriented process. Restoration practices remain deeply tied to materiality, and digital fabrication enables the convergence of thinking and making, shifting attention toward material behavior and reinforcing the tactile hands-on engagement with construction.

The introduction of new tools opens up design possibilities and intervention techniques that align with the fundamental principles of contemporary restoration. At the center of this shift is the architect-craftsperson, someone who works through digital materials to mediate between a building's historical significance and its necessary repair or integration. Through this digitally grounded approach, interventions on damaged artifacts can contribute to the ongoing stratification of history, adding a digital layer that reflects our contemporary, technologically informed experience of time. This shift ultimately redefines the role of craft. Whereas traditional craftsmanship relies on tacit, embodied knowledge of materials, digital craft demands an explicit articulation of material properties, parametric processes, and transactional relationships. As a result, fabrication workflows become a central component of architectural thinking. As designers deepen their fluency with the capabilities and constraints of these tools, new forms of expression emerge that are both materially attuned and conceptually rigorous.

European Research on the Digitization of Cultural Heritage

As a culturally diverse region, Europe carries a unique responsibility to safeguard heritage in all its forms, from art collections and historic monuments to entire landscapes. With 503 of the 1,223 designated UNESCO World Heritage Sites, the continent plays a major role in shaping global preservation efforts. In response, European institutions have expanded collective initiatives to protect cultural assets through digital tools and strategies. These efforts focus on both tangible and intangible heritage, categories defined by UNESCO to reflect the complementary dimensions of cultural identity. Digital technologies support the preservation of not only physical artifacts but also embedded knowledge and traditions. Tangible heritage includes monuments, architectural works, historic urban fabrics, handcrafted objects, and archaeological remains. Intangible heritage refers to practices and customs passed down through generations, such as oral histories, social rituals, and craft techniques. Together, they form a fabric of layered and evolving legacy, preserved through both memory and material.

Since the *2019 Declaration of Cooperation on Advancing Digitization of Cultural Heritage,* the European Union has pursued a moving target: the ambition to digitize all cultural heritage and establish a cross-disciplinary digital infrastructure to support broad access and reuse. Although progress is difficult to quantify, estimates suggest that about 30 percent of museum collections have been digitized so far. Supporting this effort is *Europeana,* a web portal created by the European Union that aggregates digitized heritage from institutions across Europe and currently provides access to over 50 million digital objects. Beyond functioning as a repository, the platform offers a framework for reuse that supports education, research, tourism, and creative industries. Metadata accompanying these objects is aggregated and standardized to enable structured search, semantic linking, and interoperability across diverse collections. Complementing this is the *Cultural Heritage Cloud,* a program focused on expanding digital resources and generating massive datasets that act as a *lingua franca* across countries and cultures. These shared datasets help refine and scale approaches to revising, rehabilitating, renovating, and restoring tangible assets. Digital information serves as input for computational workflows that can be continuously verified, updated, and reused keeping design closely tied to the conditions of time and place. Spatial computing, diagnostic tools, and digital fabrication are among the applications guiding how these strategies are implemented and advanced (Stellacci et al. 2024). However, digitization does not eliminate the risks that cultural assets face in the physical world. The *World Heritage in Danger* list, established through the *1972 World Heritage Convention in Paris*, was created to track and respond to threats such as war, neglect, climate events, and political instability. Sites ranging from historic city centers to archaeological remains are assessed based on the nature and extent of the damage, informing both immediate and long-term strategies for intervention. These actions build upon the principles outlined in international restoration charters, translating theoretical frameworks into coordinated efforts that address the material and structural realities of our cultural heritage at risk.

The two initiatives described below—*Inclusive Cultural Heritage through 3D Semantic Modeling (Inception)* and the *Time Machine Project*—represent complementary, multidisciplinary approaches to how Europe collects, preserves, and shares its cultural heritage through digital technologies. *Inception* focuses on semantically enriched 3D modeling of architectural sites, linking geometry with metadata to support conservation, interpretation, and public engagement. *The Time Machine Project* applies AI and big data to reconstruct the layered histories of entire cities, building a spatiotemporal framework that enables users to explore the past across different scales and contexts.

Inclusive Cultural Heritage in Europe through 3D Semantic Modeling

Inclusive Cultural Heritage in Europe through 3D Semantic Modelling (Inception) is a European-funded research project led by the Department of Architecture at the University of Ferrara and developed by a consortium of partners from ten countries. The project responds to the limitations of traditional methods for documenting cultural heritage, recognizing that heritage is a dynamic and evolving physical reality. To support high-quality conservation, Inception created 3D models that digitally document and semantically enrich architectural heritage. The focus was on developing

Figure 4.1 Workflow of the Inception project, illustrating the research process from data acquisition and 3D modeling to semantic enrichment and data dissemination. Diagram by Marco Medici and Federico Ferrari. Courtesy of Roberto Di Giulio.

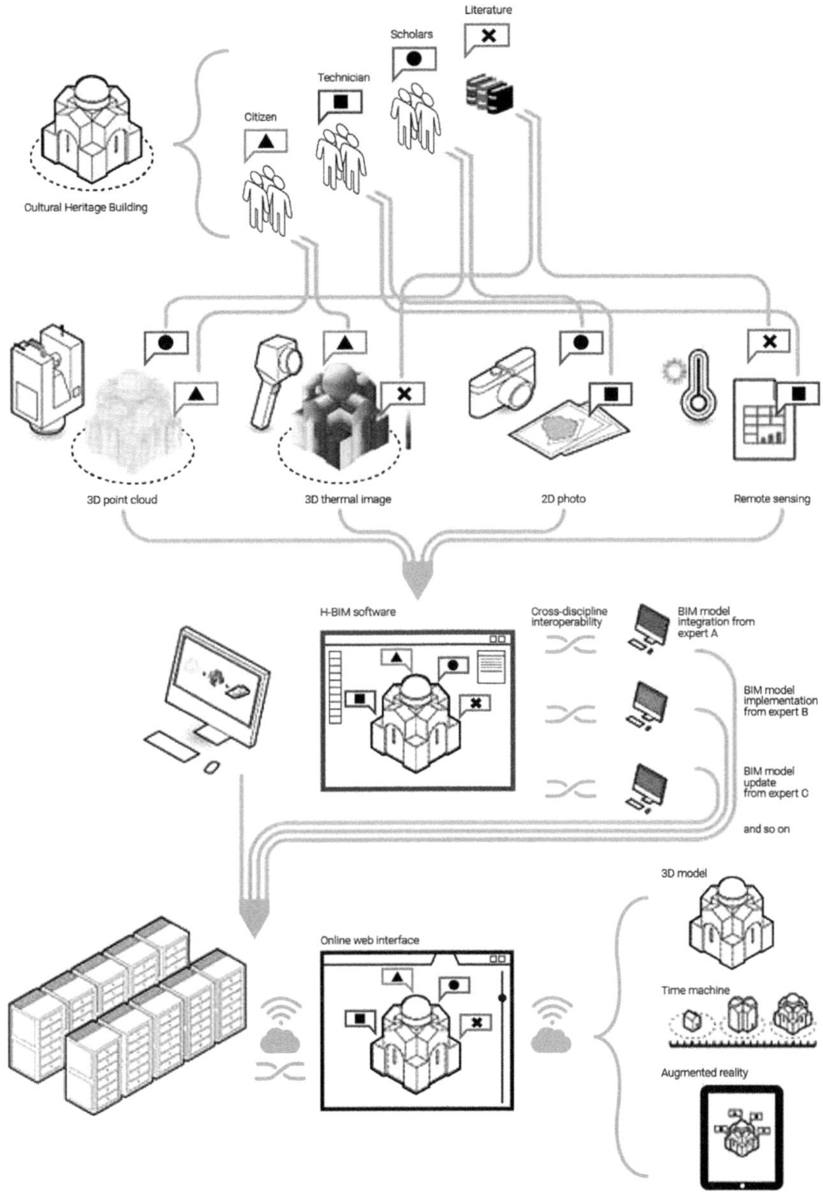

interoperable workflows (Fig. 4.1) that integrate 3D data with semantic information in Historic Building Information Modeling (HBIM) environments, improving interpretation, accessibility, and reuse. These enriched models support restoration, documentation, and public engagement and are made accessible through immersive VR platforms (Maietti, Di Giulio, Balzani, et al. 2018; Maietti et al. 2017; Di Giulio et al. 2017; Bonsma et al. 2016).

Semantic modeling refers to the process of adding structured meaning to digital data. In architecture and heritage contexts, this means connecting 3D geometry with metadata (Fig. 4.2), such as construction history, materials, techniques, or records of past damage, so that each architectural

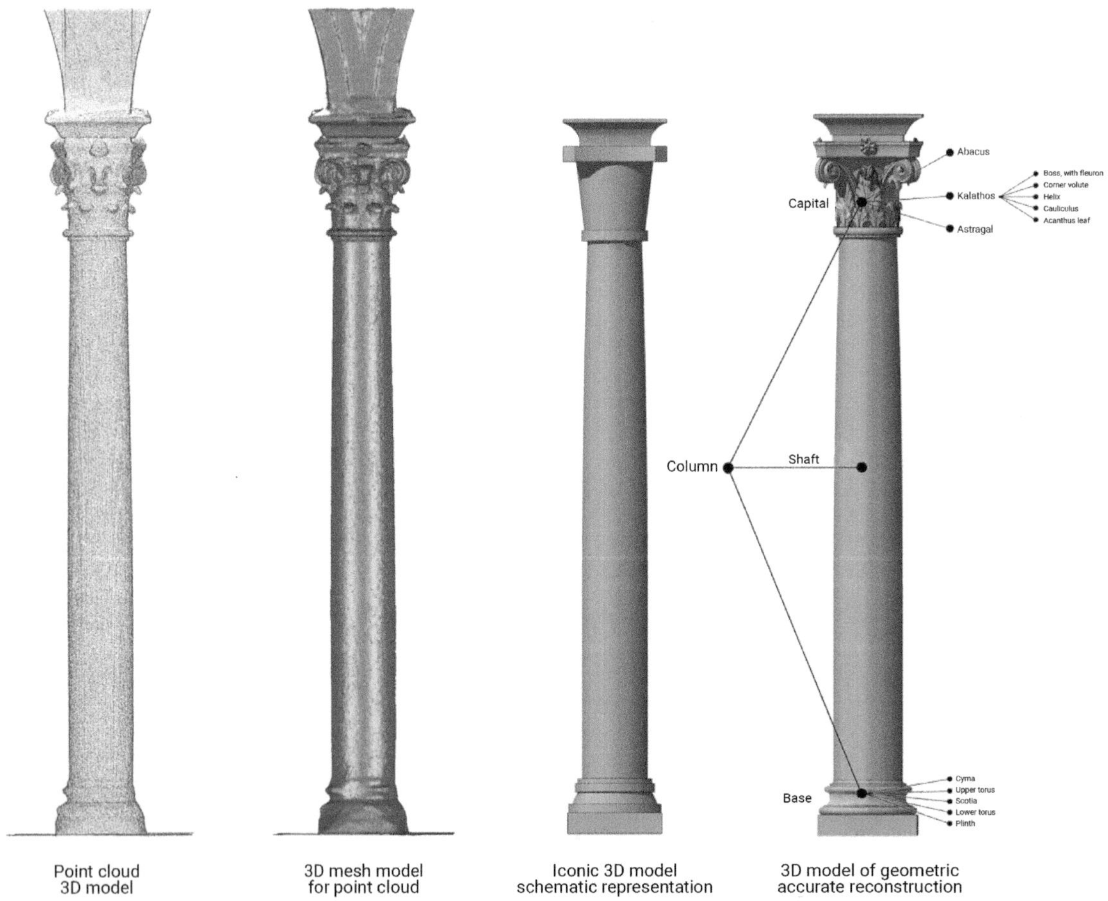

Figure 4.2 Transition from digital documentation to semantic modeling. All information is structured as semantic triples, where two elements—a subject and an object—are connected by a predicate that defines their relationship. The existing building is thus broken down into elements, each associated with specific properties and linked to other elements through physical or functional relationships. Courtesy of Roberto Di Giulio.

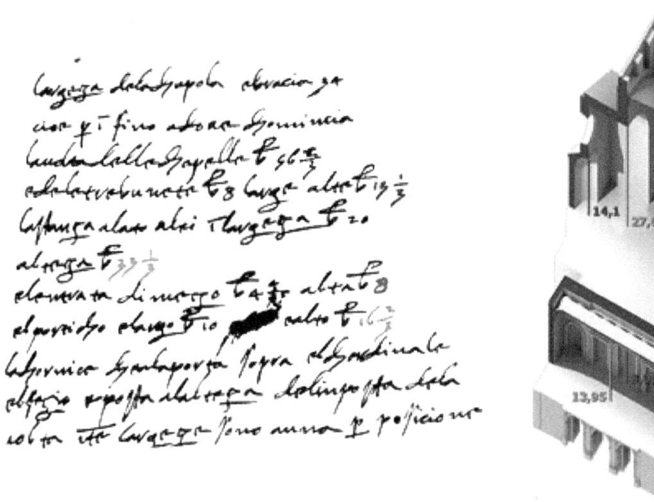

Figure 4.3 Detail of the BIM model of San Sebastiano Church in Mantua, designed by Leon Battista Alberti. Generated from point cloud data, the model serves as a graphical verification of measurements based on early 16th-century drawing by Antonio Labacco (architect, engraver, and writer, pupil of Antonio da Sangallo). Courtesy of DIAPReM. The drawing is cited by the Inception team in the paper *Digital Memory and Integrated Data Capturing* in *Mixed Reality and Gamification for Cultural Heritage*.

element is not just seen, but also understood beyond representation with contextual insights. Ontologies[3] provide the underlying conceptual structure for semantic modeling by defining key concepts and their relationships. This approach to integrating information is foundational to HBIM, which adapts BIM methodologies to the documentation and care of historic architecture (Logothetis et al. 2015). HBIM combines accurate 3D geometric models, often created through photogrammetry or laser scanning (Fig. 4.3), with semantic data such as construction history, materials, previous interventions, and conservation needs. The result is a comprehensive digital framework that captures not only a building's current form but also its transformations over time and the knowledge required for its stewardship. Unlike conventional BIM, which is forward-looking and used primarily in new construction, HBIM is retrospective and interpretive, designed to address the layered complexities of heritage sites.

This structured understanding of architectural elements has deep roots in Western architectural theory. Since antiquity, thinkers have sought to codify the "grammar" of architecture, the rules and relationships that govern composition, proportion, and form. Foundational texts such as *De Architectura* by Vitruvius, *De Re Aedificatoria* by Leon Battista Alberti, and *I Quattro Libri dell'Architettura* by Andrea Palladio established enduring frameworks for classifying built form and understanding its logic. These treatises were early attempts to systematize architectural knowledge, grounded in the desire to make meaning explicit, transferable, and applicable across contexts. Theoretical developments in the late 20th century continued this trajectory. In *The Logic of Architecture*, Bill Mitchell introduced the idea that architectural design could be articulated through formal languages, comparable to logical grammars. His approach viewed design as a computational process, where rules and relationships generate spatial outcomes (Mitchell 1998). Mitchell's emphasis on structured, symbolic systems aligns closely with the principles of semantic modeling and HBIM, both of which aim to transform geometric models into intelligent, context-rich representations.

Inception demonstrates that cultural heritage is not just a collection of geometric data, but a layered repository of historical, material, and symbolic information. This includes the names and classifications of building components, decorative elements, alterations over time, signs of deterioration, and documents outlining potential future risks to structural integrity (Maietti, Medici, et al. 2018). When layered onto quantitative survey data, a more complete and intelligible digital representation is created. Each building embodies diverse solutions to practical needs, shaped by additions and modifications accumulated across generations. In recognition of this complexity, the *Inception* project introduced shared parameters and a standardized glossary of architectural terms as a foundation for semantic enrichment within BIM environments. Identifying and classifying individual architectural components requires a common language, which this glossary helps to establish. To broaden the impact of its work, the project used VR to present enriched 3D models to a wide public (Maietti, Di Giulio, Piaia, et al. 2018). Interoperable data can also construct predictive models that can trace the likely evolution of building degradation. Semantic models linked to this type of algorithmic logic could eventually inform strategies for intervention, maintenance, and preservation planning, especially when paired with the development of tools designed to support such workflows.

Time Machine Project: Processing the Big Data of the Past

The *Time Machine Project* originates from the ambition to link cultural heritage with its wider social and historical context, visualizing connections across time and space. The project promotes a multidimensional understanding of heritage, moving beyond digital 3D reconstructions and the simple archiving of information. Developed through a collaboration between the École Polytechnique Fédérale de Lausanne (EPFL) and Ca' Foscari University of Venice, the effort focuses on building big data of the past: a massive, searchable digital system that maps Europe's social, cultural, and geographical history across centuries. Its main application is the *Venice Time Machine*, launched in 2012 and still ongoing. The work consists of the historic archives of Venice to construct a multidimensional model of the city's cultural heritage (Fig. 4.4), spanning more than 1,000 years (di Lenardo and Kaplan 2015). This mass digitization effort targets the *Archivio di Stato*, one of the largest archives in Venice, a national system that preserves, manages, and provides access to the documentary heritage produced by the state and public institutions over time. It is estimated to hold 80 kilometers, or approximately 50 miles, of shelves filled with administrative documents. Through this multidimensional model, researchers are reconstructing the history of Venice in both 2D and 3D for any point in time, from the city's origins to the present day incorporating any event that left a written trace (Karch et al. 2025). The model draws on a wide range of documents, including economic transactions, population censuses, public works records, news circulation, commercial activity, migration data, birth and death registers, tax statements, maps, urban planning schemes, as well as artistic and architectural drawings. All documents are handwritten, primarily in Latin or the Venetian dialect, and span from the medieval period to the 20th century.

The complexity of interpreting these historical materials is addressed through machine learning systems. Text recognition algorithms identify and associate letter shapes, even when fonts vary, allowing handwritten content to become searchable and cross-referenced across vast collections. *Time Machine's* algorithms are developed to analyze the structure

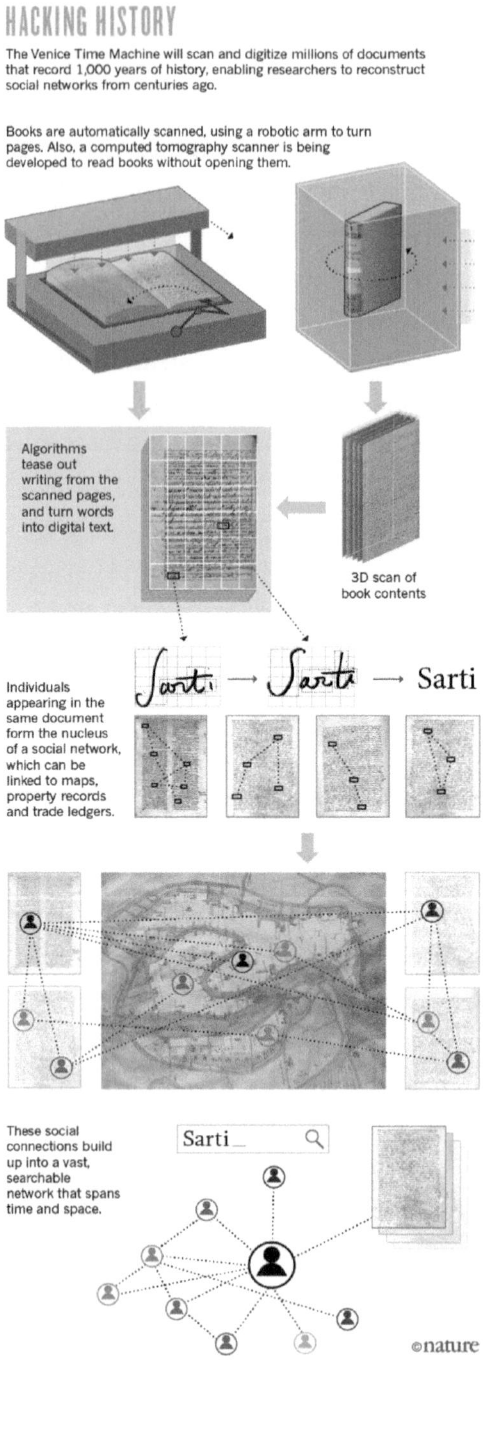

Figure 4.4 The *Time Machine* workflow: from mass digitization and AI transcription to semantic enrichment and spatial mapping, enabling historical reconstruction and cross-disciplinary research. Courtesy of Frédéric Kaplan and the Time Machine research team.

of written text and detect recurring graphical forms, enabling connections between similar elements. This process is supported by READ (Recognition and Enrichment of Archival Documents), a technology designed for the automated transcription, recognition, and semantic enrichment of historical handwritten sources. For particularly fragile manuscripts, robotic arms carefully turn pages in controlled environments, while tomography scanners, adapted from medical imaging techniques that construct 3D images from X-rays taken at multiple angles, are being developed to reveal content without physically opening the documents.

By combining information and converting handwritten documents into digital, searchable text, researchers can reconstruct large segments of Venice's past and uncover previously hidden narratives—complete biographies, political dynamics, or even the appearance of buildings and entire neighborhoods. The *Venice Time Machine* also links citizens and researchers to historical cartographic records, including 16th-century maps that depict the city's urban fabric. The use of AI in historical research has led to the development of algorithms capable of recognizing shapes in old drawings and performing transversal searches, which query across multiple data types, domains, or sources simultaneously. This makes it possible to trace, for example, all the economic transactions involving a person who commissioned a work, or to identify all architectural drawings featuring pitched roofs. Information extracted from primary sources is structured into a semantic graph of linked data and visualized through a historical geographic information system, forming the most extensive database ever created for Venetian documents (Abbott 2017).

Multidisciplinary collaboration has been essential to the development of the *Time Machine* workflow, which brings together 3D modeling, big data, AI, and extended reality. The initiative aspires to create a pan-European infrastructure capable of digitizing and analyzing historical data from multiple cities using the same research methodology (Fig. 4.5). The project also aligns with broader cultural heritage goals, as reflected in its manifesto, which echoes the *2019 Declaration on the Digitization of Cultural Heritage* and its call for the strategic use of digital tools to document,

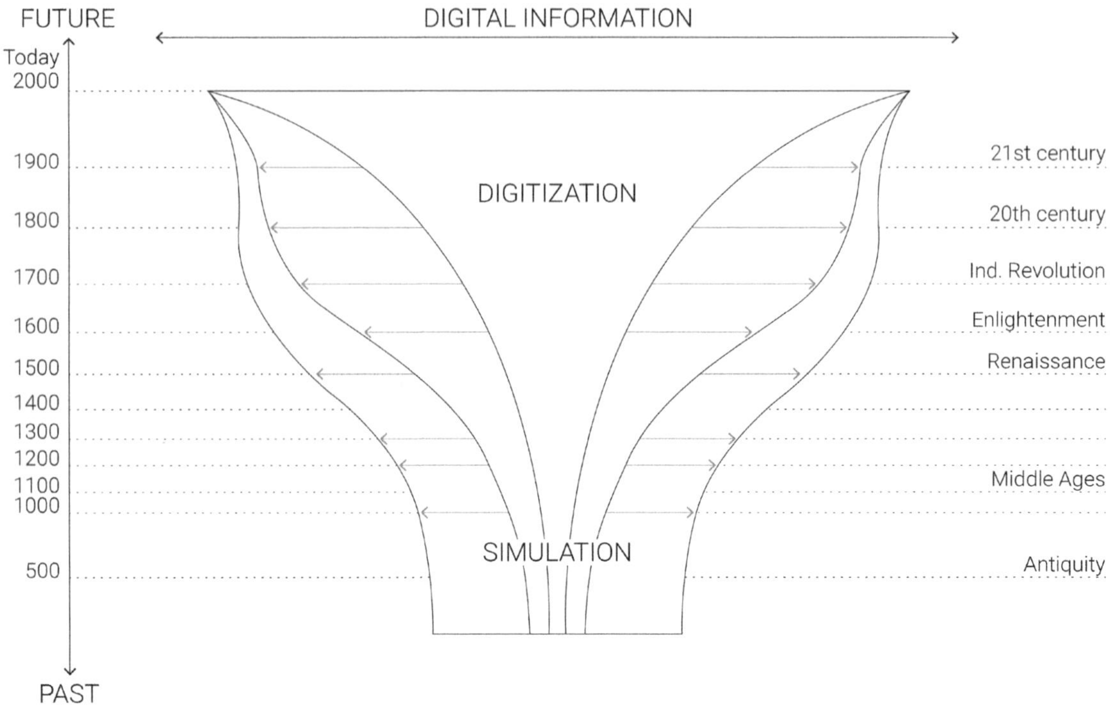

Figure 4.5 The Information Funnel: as the curve recedes into the past, it narrows, illustrating the decreasing availability of data. Reconstructing a past as detailed as the present requires both extensive digitization and data extrapolation through simulation. Source: Frederic Kaplan speech at TEDx Lake Geneva.

preserve, and improve access to cultural assets. The approach combines large-scale archival digitization with advanced computational techniques to produce searchable, re-constructible, and analyzable historical models, ushering in a new phase of digital humanities and expanding the possibilities for understanding human culture through the lens of the digital age.

Digital Approach to Historical Languages

Historically, nothing in our past was created digitally. The physical world has always been shaped and manipulated by human *digiti*, our fingers, through direct engagement with materials and the interpretation of various notations or instructions, from study models to hand-drawn diagrams. The hand has long played a central role in learning, anchoring design thinking in the physical realm and functioning, in many ways, as an analogue tool for translating ideas into form (Pallasmaa 2009). Within this framework, architecture and other creative practices rely on tacit knowledge, a form of understanding that cannot be codified but is instead cultivated through embodied experience. Craft traditions embody this principle, privileging the learned environment by learning through making over only theoretical instruction. While digital tools attempt to bridge the gap between concept and execution, they have also raised concerns about a growing disconnection between designer and material. Digital design often unfolds in a disembodied way, where touch and physical interaction are absent (Codarin and Daubmann 2022). When this tactile dimension is removed, projects risk losing the richness and nuance that come from hands-on engagement. This raises a pressing question: how can we reintroduce a substantiated, human sensibility to the restoration site, when contemporary modes of architectural conception are increasingly removed from the material practices that once defined them?

At the same time, the emergence of advanced software tools rooted in calculus, an evolution from Euclidean geometry, has opened new formal possibilities for design. Designers can now produce complex, deeply intricated geometries and responsive structures through digital sculpting, topological optimization, subdivision modeling, physical simulation, and procedural scripting—operations that far exceed what can be achieved using traditional geometric logic. Digital technologies offer an opportunity to interpret historical forms, developing new visual languages that resonate with the past while introducing aesthetic and conceptual shifts. Within this exploration, precision itself becomes negotiable, recalling the kind of human imprecision that once produced artifacts in the physical world. Computational tools reveal space for alternative readings of form, where slight deviations or digital "glitches" may suggest unexpected continuities between digital processes and the expressive imperfections of human craft. These moments of ambiguity open possibilities for reimagining how technology participates in shaping meaning—allowing designers to navigate the space between codified representation, excessive resolution, and human expression, venturing into new territories.

The challenge, then, is how to structure a method for integrating new tools, languages, machines, and fabrication processes into a field historically grounded in craft, while also addressing how to digitize the tacit and manual knowledge of craftspeople. Can a computational understanding of history propose new approaches to restoration in which precision, repeatability, and performance align with an evolving aesthetic sensibility? And can digital workflows offer methods of cultural engagement that go beyond preservation to enable transformation? The projects described in the following sections present a series of experiments in digital design and fabrication, exploring how cultural knowledge, both historical and speculative, can be reproduced, interpreted, and conserved. These works establish connections between tradition and innovation, investigating some converging and complementary ways to engage with the past through the lens of contemporary design.

Digital Ornament

Since the early 2000s, an expanding body of literature has addressed the return or reconfiguration of ornament in architecture. Mario Carpo describes ornament's fluctuating status across history, noting that it has been perceived at times as a valuable supplement and at others as waste (Carpo 2017). In architectural theory, the roots of aesthetic reasoning can be traced back to Vitruvius and Alberti, who both emphasized the inseparability of *utilitas* and *venustas*—usefulness and beauty—in what we now recognize as historic architecture (Di Stefano 2010). This relationship is clearly embodied in Rome's enduring monuments such as the Pantheon and the Roman Forum. Their longevity—one still functioning as a church, the other preserved as an archaeological site—is not solely due to material performance over the centuries, but also credited to human efforts that preserved their intrinsic beauty by aligning purpose, structure, and materiality. When these elements are in balance, ornament intensifies the architectural experience by contributing a symbolic and emotional resonance.

Adolf Loos's 1913 essay *Ornament and Crime* presents a modern critique that links ornament to waste and inefficiency. He argued that removing decoration would reduce unnecessary labor and conserve resources. Yet this raises a critical question: did the rejection of ornament truly benefit workers, or did it simply shift the nature of their tasks? Today, artificial intelligence and digital fabrication offer a different lens for exploring this issue. These tools allow ornament to re-emerge through computational processes, reframing labor, cost, and material use. In this setting, ornament reflects a new kind of subjectivity specific to digital culture that no longer depends on human craftsmanship alone or bears the moral and economic pressures associated with traditional labor (Lee and Macgillivray 2023; Cayer 2018). One project that directly engages Loos's critique is *Procedural Sinters*, by Andreas Körner, which begins with a reflection on Loos's conflicted relationship with ornament, and particularly his use of sliced stone as veneer surfacing despite his public opposition to decorative excess. The project takes Loos's incrustation[4] finishes as both conceptual provocation and visual input, using imagery from his decorative stonework as a dataset. These patterns are extended through machine learning, which generates new ornamental surfaces by interpreting and evolving the original material logic. The results are parametrized into 3D geometries that propose a digitally reimagined materiality, reworking the set of learned patterns into novel formal expressions (Fig. 4.6). The physical prototypes challenge Loos's legacy by positioning ornament not as aesthetic excess, but as a procedural, expressive, and environmentally attuned architectural language (Körner 2024).

In everyday usage, ornament is often understood as decoration that lacks practical function. However, ornament also carries historical and symbolic meaning as it helps situate objects to identify time and context, revealing cultural values. In this sense, ornament becomes a mechanism through which architecture connects to broader cultural narratives, mediating between material presence and social meaning. Marjan Colletti explores ornament as something expressive and exuberant, shaped by computational tools that stretch, layer, deform, and generate dynamic surface articulations. Ornament, in this context, reflects a broader condition in contemporary architecture, where the discipline has gained fluency

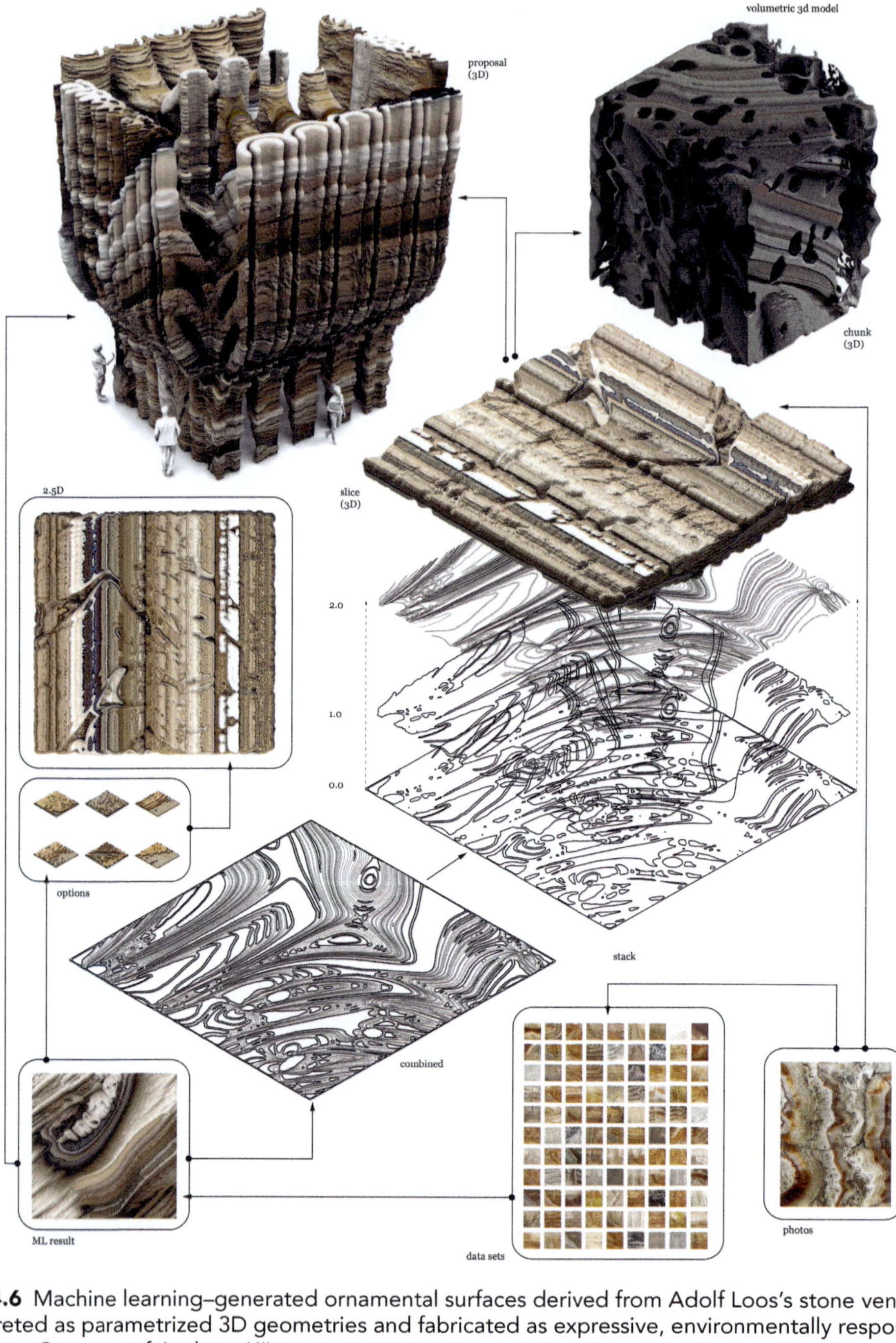

Figure 4.6 Machine learning–generated ornamental surfaces derived from Adolf Loos's stone veneers, reinterpreted as parametrized 3D geometries and fabricated as expressive, environmentally responsive prototypes. Courtesy of Andreas Körner.

with parametric modeling and three-dimensional software (Colletti 2010). This technical proficiency has allowed for a new kind of expression, where surfaces and volumes are shaped with intensity and richness that were previously unattainable. The result is an architecture defined by ornate and exuberant articulations that convey complexity, precision, and elegance through digital means.

Similarly, Antoine Picon views ornament as a sensory and effective dimension of architecture, tied to evolving definitions of materiality. For Picon, ornament is not only a matter of form but also of experience, reflecting how matter is shaped, perceived, and given meaning through cultural and technological contexts (Picon, 2014). This expanded understanding also includes the marks left by digital fabrication processes. Toolpaths from CNC routing or robotic milling, for instance, can be interpreted as a form of ornament in themselves, a visual and tactile byproduct of making that becomes part of the surface expression (Lynn 2004). This conversation extends into projects such as *Ancient Future*, an installation connected to a live performance in which a carved beam is collaboratively produced by both craftspeople and a robot guided by artificial intelligence. The artisan directs the creative vision, while the robot carries out tasks requiring precision and speed. The project frames robotics not as a replacement for manual labor but as a complement to it, enhancing traditional practices and allowing artisans to dedicate more energy to expression itself (Fig. 4.7).

The return of ornament can be traced not only within built architecture but also in creative research and critical historiography, where its relationship to cultural and technological conditions is examined. This line of inquiry is explored in academic coursework, such as *Intense 3D Printed Ornament,* taught by Riley Studebaker at Rensselaer Polytechnic Institute's School of Architecture, where students scan historical ornament and develop workflows that generate new physical artifacts (Fig. 4.8). This renewed presence emerges in the design of surfaces that resist neutrality, moving beyond the "featureless" aesthetic to convey intention, depth, and complexity. The ability of ornament to respond to context and incorporate historical reference as both structural and conceptual grounding

Figure 4.7 *Ancient Future* project at 2025 Venice Architecture Biennale. The installation demonstrates human-robot collaboration in the creation of highly detailed ornament. Project by Bjarke Ingels Group, Laurian Ghinitoiu, and Arata Mori. Photo by Sara Codarin.

establishes a meaningful continuity between past and present. Within digital culture, ornament is also a mode of communication—at times direct, at other times ambiguous—but consistently rich with meaning. Rather than suggesting a nostalgic return to the past, ornament today amplifies the evolving culture of digital materiality and portrays technology as a cultural force. In this capacity, ornament holds the potential to re-enter the field of conservation, introducing a contemporary lens through which to reinterpret heritage.

Digital Baroque

A constructive starting point for structuring a new aesthetic of restoration can be found in the work of scholars who treat historical interpretation as a foundation for future design. There is a current fascination with Baroque architecture that stems from both cultural and technological shifts: on the one hand, renewed discourse on ornament and historical references has reopened its relevance within design theory; on the other, advances in

Figure 4.8 *Printer Formwork*, 2024, a temporary sculptural installation designed by Aiden Olsted under the guidance of Riley Studebaker at Rensselaer Polytechnic Institute's School of Architecture. Installed at Paine Castle in Troy, NY. Courtesy of Riley Studebaker.

high-resolution imaging and data capture technologies provide unprecedented access to the complexity of Baroque spatial constructs. These tools reveal the richness of its forms with new digital dimensional clarity, positioning the Baroque as a territory for contemporary architectural exploration.

Marjan Colletti's work draws from Baroque architecture and culture as a space for hybridizing digital, analog, biological, and artificial media. In his research, computational design and digital fabrication are presented as agents that reshape understandings of beauty, reality, and design within what he describes as the post-digital condition. Colletti extends this thinking toward a neo-Baroque sensibility, suggesting that contemporary architecture resonates more with Baroque values than with modernist ones. This shift is both aesthetic and cultural, reflecting a broader orientation toward complexity, ambiguity, and spectacle. Digital tools now enable detailed control over form and texture, echoing Baroque ornamentation where 3D and 4D design methods can replace flat, static representations with immersive, dynamic spatial experiences. Robotic systems are increasingly involved in both fabrication and conceptual development, supporting geometries and material behaviors that emphasize performance and transformation. These developments encourage open-ended, a-tectonic compositions and interdisciplinary approaches, moving away from modernist clarity and modularity. Gradients, morphing forms, and ambiguous spatial conditions support a responsive mode of design that aligns with contemporary restoration practices shaped by technological and cultural complexity (Colletti 2017).

The post-digital neo-Baroque work led by Marjan Colletti unfolds across a three-volume publication. It begins by establishing a conceptual bridge between Baroque architecture, postmodern theory, and digital culture (Colletti et al. 2024; Colletti 2024; Colletti and Massin 2024). This foundation is expanded through experimental projects that use machine learning, 3D printing, and computational design to reinterpret ornament, materiality, and natural systems. The series also incorporates interdisciplinary perspectives that highlight the shared aesthetic of this evolving design language, marked by the collapse of binaries between the natural and the artificial, along with the digital and the physical. Nature plays a central role in this discourse, not as a direct source of imitation but as a procedural model for generating form. Complexity, transformation, and layering—qualities inherent to natural systems—are translated into digital

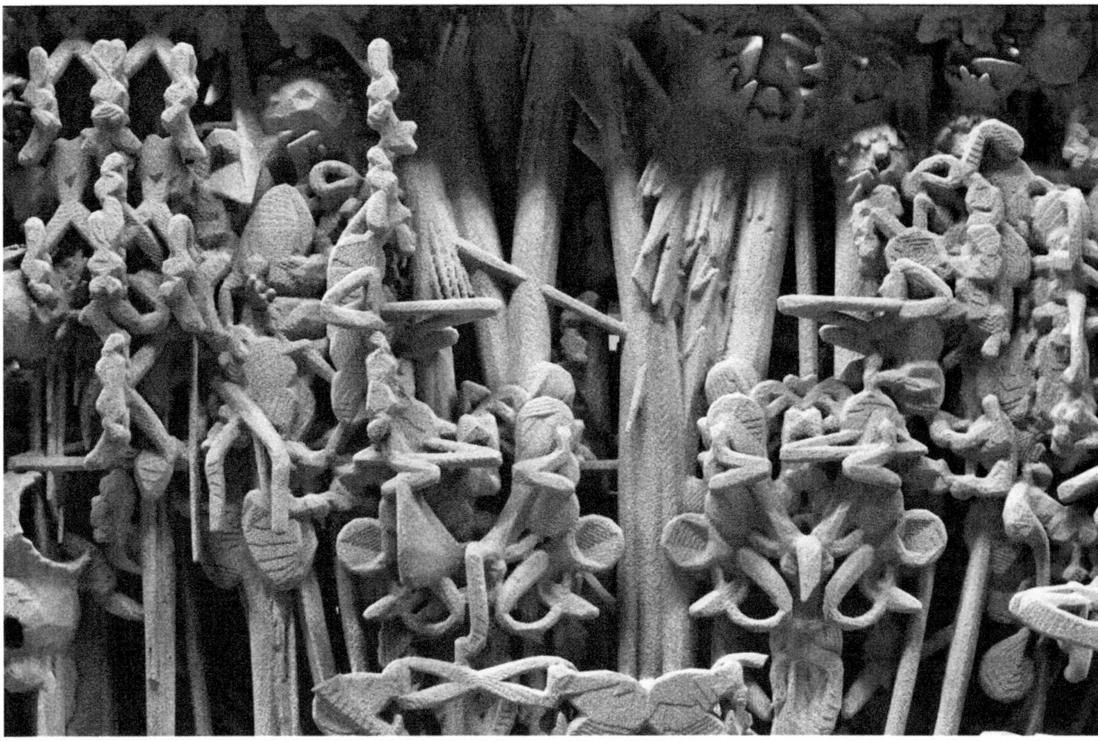

Figure 4.9 *Plantolith*, 2013. 3D silica sand print, approximately 250kg, or 551 lb. Courtesy of Marjan Colletti.

design logics, where geometry emerges from code and algorithms. One manifestation of this design sensibility is the 2013 prototype *Plantolith*, a built form that merges the opposing formal logics of plants and monoliths through large-format 3D printing (Fig. 4.9). *Plantolith* simulates natural processes, blurring the line between organic growth and tectonic structure. Its components are digitally unified to produce a uni-material monolith made of silica sand, fabricated through an additive manufacturing process that chemically binds material layer by layer into its final form.

A research-based reference for Baroque spatial systems is *Rules of Tools*, a project conducted by Karl Daubmann at the American Academy in Rome. The work explores historical and tectonic rule sets to parametrically engage with contemporary material and technological conditions. Drawing from the legacy of design-builder Francesco Borromini, the project

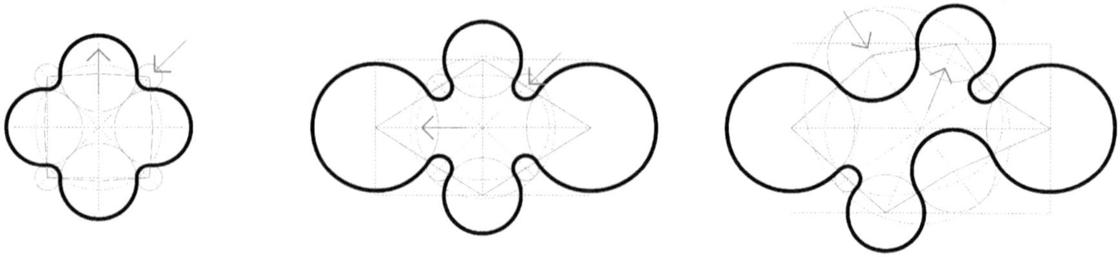

Figure 4.10 Relaxation of symmetry rules in a quatrefoil geometry, a motif frequently employed by Borromini. Courtesy of Karl Daubmann.

investigates how novel spatial arrangements were achieved in Baroque architecture through specific materials and construction techniques. Borromini's subtractive method, based primarily on stone cutting, enabled the creation of flowing forms that concealed individual building elements within continuous surfaces (Daubmann and Jacobi 2019). Building on this lineage, *Rules of Tools* examined Baroque churches alongside Borromini's drawings, using them as a reference to initiate parametric plans that translate historical design logic into computational form (Fig. 4.10). Digital models were developed using procedural operations to augment spatial complexity through incremental steps. This method reflects Borromini's drawing techniques, which often are comprised of simultaneous variations. The models, developed in CATIA, function as geometric scripts composed of simple sequences and were visually validated against the actual spaces. Fragments were 3D printed and incorporated into the drawings to inform formal understanding through materialized geometry (Fig. 4.11). A parallel line of analysis focused on bodily experience, involving on-site observation, photography, and attention to ritual and movement within space (Fig. 4.12). The repetition of these bodily gestures was conceived as something that could eventually be programmed for robotic execution (Daubmann 2021), with parameters such as speed, duration, complexity, and path imagined as variables. One might also imagine correlations between gesture and space, the play of light, and the spatial detail revealed in each image (Fig. 4.13).

From the study of Baroque architecture, creative opportunities emerge in the work of Andrew Saunders. In the research project *Baroque*

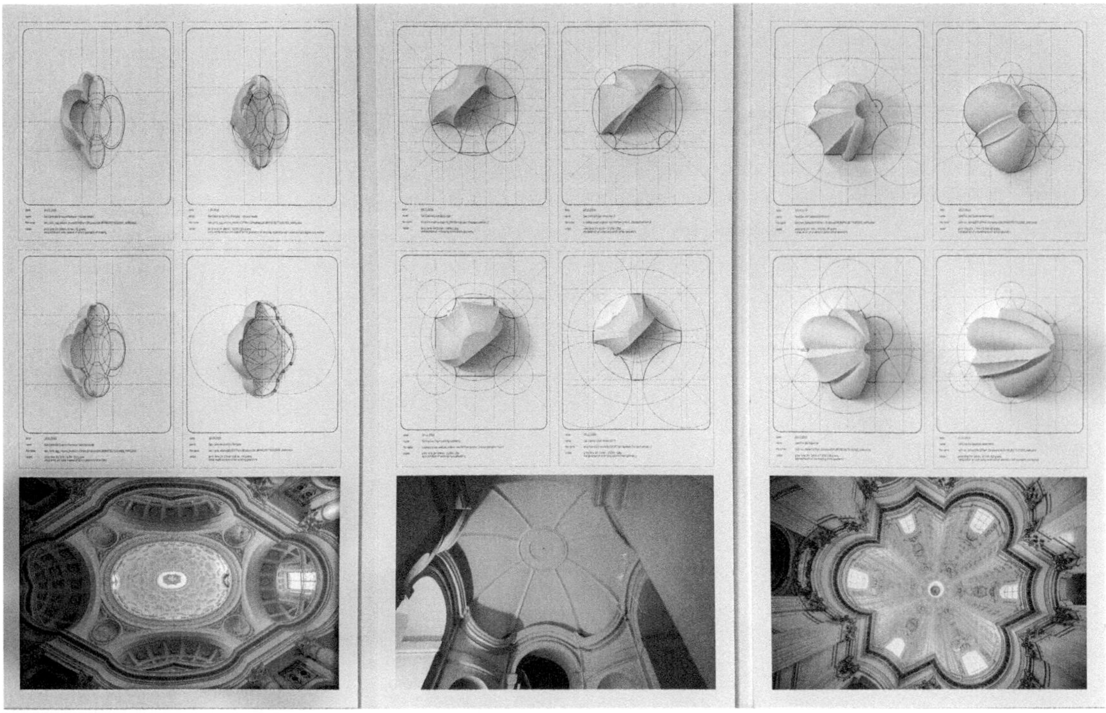

Figure 4.11 Studies of Borromini's domes, 2015–16. Defining and relaxing procedural rules for the construction of space. Courtesy of Karl Daubmann.

Topologies, Saunders reinterprets Baroque form through digital design principles rooted in topological thinking (Fig. 4.14). Transformation replaces static form as the primary driver of spatial expression, and ornament becomes integral to the surface, embedded within its geometry. Curved geometries and spatial distortions evoke dynamic visual and bodily perception, echoing Borromini's spatial experimentation. Through digital simulations, material behaviors are modeled to explore how form and structure might evolve together under virtual forces. The resulting digital constructs are subsequently 3D printed, translating the simulated geometries into tangible prototypes that test how continuous surfaces can merge ornamental and structural roles. These design strategies extend Baroque complexity into a contemporary architectural language shaped by computation and formal fluidity (Saunders 2018; 2017; 2009). This research

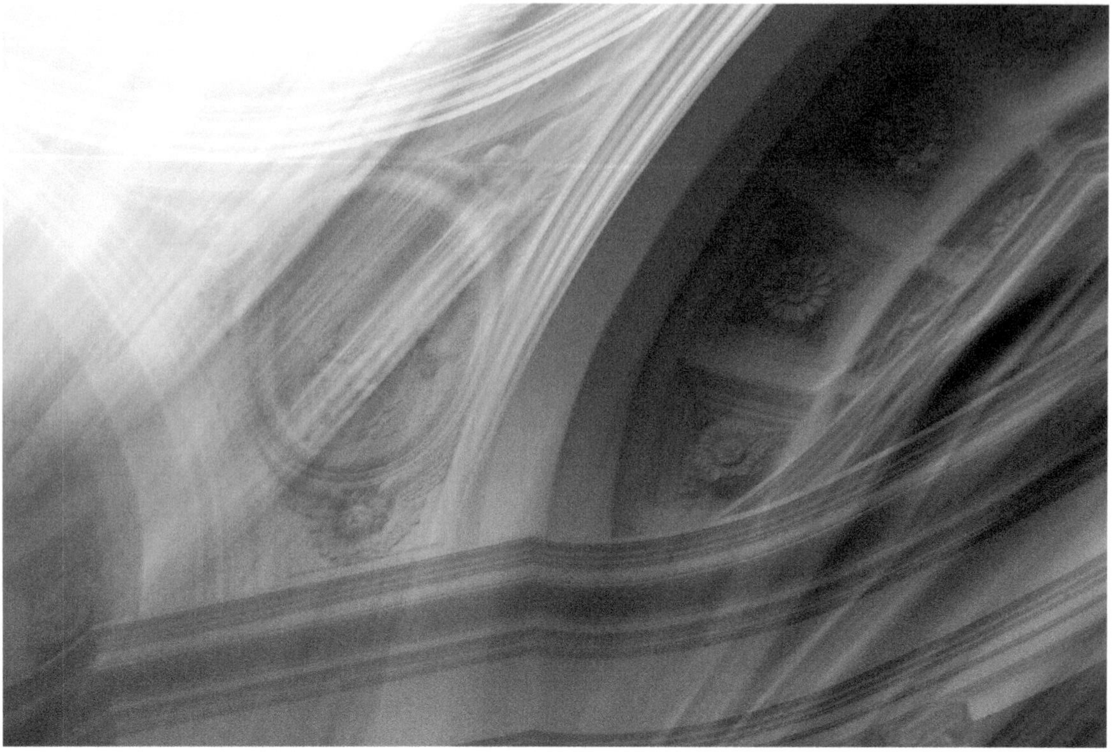

Figure 4.12 Long-exposure time-lapse photograph of Borromini's San Carlo alle Quattro Fontane, 2015. Courtesy of Karl Daubmann.

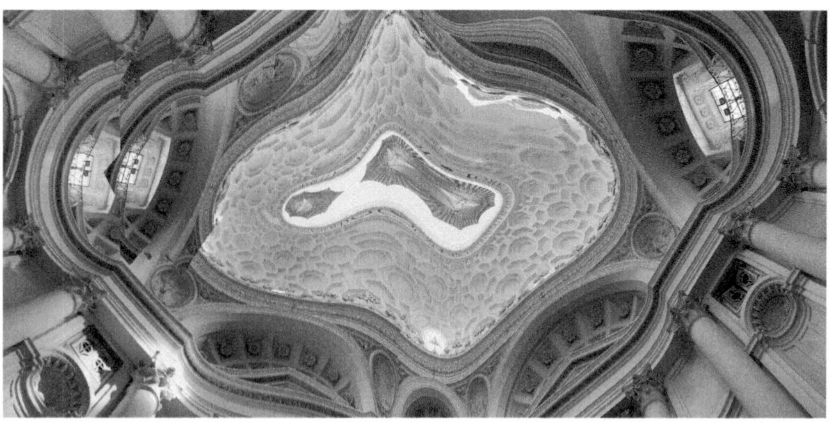

Figure 4.13 Distorted panoramic photograph of San Carlo alle Quattro Fontane. The warped geometry evokes the visual logic of an early AI-generated spatial reconstruction. Photo by Sara Codarin.

Figure 4.14 *Spheroidal Cosmologies* series: San Carlo alle Quattro Fontane from *Baroque Topologies* by Andrew Saunders, 2018.

also informs Saunders's elective seminar *Pointwork Paintwork* taught at the University of Pennsylvania, which builds on Baroque scholarship and the history of architectural representation. Students engage directly with LiDAR scans of canonical churches and are challenged to work exclusively with point clouds, moving away from conventional vector-based or compositional analysis. Through this technique, spatial features and recurring design tropes emerge from dense, dynamic fields of points (Fig. 4.15). Variations in density, direction, and color produce expressive

Figure 4.15 *Pointwork Paintwork* series: San Carlo alle Quattro Fontane main altar (left) cornice (center), interior elements (right) by David Perrine from the *Baroque Topologies* seminar taught by Andrew Saunders, 2023.

representational effects, offering a more painterly, textural, and colorful reading of the Baroque interior. These data-rich, time-based representations reframe the Baroque not only as a historical subject but as an active spatial and visual inquiry.

The presented works collectively demonstrate how digital Baroque can serve as a productive framework for developing a credible new aesthetic, engaging with monumental architecture through advanced tools of analysis. By working with embedded geometries, material behaviors, and construction logics—we can trace a lineage from historical master builders to contemporary digital practitioners and these projects reveal how the intersection of this historical knowledge, representational innovation, and digital fabrication can reframe restoration as an active design practice that is both culturally grounded and projective.

Speculative Artifacts

It would be mistaken to assume that restoration through transformative design is either easy to accomplish or widely practiced. In practice, most real-life restoration efforts remain conservative, typically limited to cleaning, rehabilitation, and structural consolidation—interventions in which any new material trace must remain recognizable, or, as Jorge Otero-Pailos puts it, must be "confessed" to historians (Otero-Pailos et al. 2016). At the same time, it is precisely in this context that it is productive to explore design strategies that forge connections not yet established, particularly between digital tools and emerging aesthetic approaches that fall outside traditional restoration frameworks but hold growing relevance within its discourse.

The following speculative projects are grounded in research processes made explicit through material experimentation and physical prototyping; they draw upon cultural references from our contemporary phenomenological world[5] and carry the potential to shape a narrative still in formation, connecting past and present while contributing to a meaningful dialogue.

Including such work also acknowledges that speculative objects, typically associated with artistic production, might, or perhaps should, have a role in the discourse of preservation. Traditionally, this realm has been dominated by architects, planners, social scientists, historians, and conservation experts, but there is value in making space for speculative aesthetic inquiry as a legitimate form of knowledge production. Such inquiry can expand how we engage with memory, material value, and cultural continuity.

In many of these projects, the formal language conveys a historical resonance, opening up new possibilities for engaging existing buildings through strategies of integration, addition, repair, or even the reconstruction—or "re-edition"—of what might be considered a "lost text" (*riedizione del testo perduto*). This design approach may particularly resonate with designers who, as Manfredo Tafuri observed, are by nature inclined to leave a mark on the built environment—unlike conservation professionals, who in his view are defined by their intent to leave no trace of themselves in the physical fabric of what already exists (Tafuri 1991).

Hyperobjects

One aspect of transmitting built heritage to future generations lies in material engagement. Every intervention becomes a trace of the time in which it was made, a material index of contemporary values, technologies, and worldviews. Work on existing architecture requires establishing a relationship with the material evidence of the past. This engagement needs to be accompanied by a critical assessment of the cultural value of the artifact (*giudizio di valore*). When material interventions are necessary, Francesco Dal Co, drawing on the theories of Alois Riegl, argues that appreciation for the antiquity of a monument must be preceded by historical evaluation in order to meaningfully align with the value of novelty (Dal Co 2013). Whether addressing the reintegration of volumetric losses or focusing on maintenance and conservation, restoration inevitably entails—however subtly—material and formal transformations. As such, it calls for an interpretive approach capable of resolving the "architectural text," where layered readings clarify the project's material and formal logic. In this context, "text" refers to the grammar, syntax, and semantic depth of a building, understood as a legible whole composed of interrelated parts (Dalla Negra 2017).

Extending this notion of restoration as an interpretive act, what does it mean to activate, re-functionalize, or make more performative existing architectural heritage in an era when humans are intensely involved in the production of anthropogenic systemically entangled phenomena? These phenomena, also defined by Timothy Morton as hyperobjects, are digital/physical entities that result from human manipulation of natural systems at a planetary scale, such as the internet, climate change, or even the geological imprint of human activity over deep time (Morton 2013). Hyperobjects are massively distributed across time and space; they emerge from technological, industrial, or ecological processes and exceed individual perception and comprehension. Their emergence coincides with what Luciano Floridi describes as the transition into hyperhistory, a historical phase in which the production, management, and transmission of information have become infrastructural (Floridi 2015), shaping the way we understand and intervene in the physical world.

Architectural heritage, then, becomes a node within an informational ecosystem, situated within broader digital, ecological, and cultural entanglements. When brought into dialogue with restoration theory, hyperobjects are brought to the human scale and compel a rethinking of how we approach historical architecture—not as static artifacts to be preserved in isolation, but as participants in larger ecological, social, and technological networks. Traditional restoration often seeks to return a structure to an idealized state, guided by historical, critical, and aesthetic judgments. Yet the presence of hyperobjects complicates this approach. The material decay of stone accelerated by acid rain, the evolving energy performance demands of climate adaptation, or the embedded carbon of past construction methods are not background conditions; they are active agents expanding the scope of restoration today. In this broader frame, restoration becomes less about recovering a cultural artifact and more about negotiating with long-duration forces where adaptive practice must operate with an awareness of planetary responsibilities alongside various interdependencies.

A parallel can be made with the creative logic of *Digital Grotesque II*, a project that operates at the intersection of materiality, memory, and time. *Digital Grotesque II* reimagines the grotto—an ancestral, primordial interior—through algorithmic processes and additive manufacturing (Fig. 4.16). The project, developed by Benjamin Dillenburger and

Figure 4.16 *Digital Grotesque II*, 2017, a full-scale 3D printed sandstone grotto exploring algorithmic ornamentation and immersive spatial complexity. Courtesy of Benjamin Dillenburger and Michael Hansmeyer.

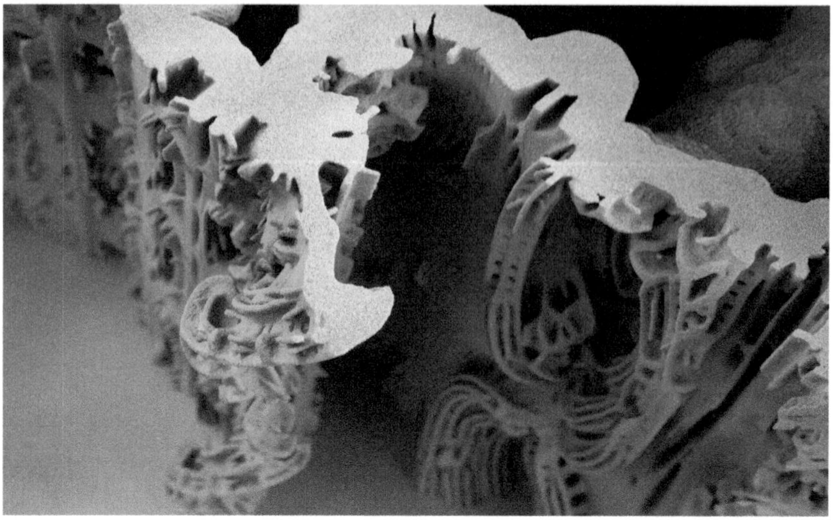

Figure 4.17 *Digital Grotesque II.* Ornamentation detail and section cut. Courtesy of Benjamin Dillenburger and Michael Hansmeyer.

Michael Hansmeyer, results in a structure that resists binary classifications: it is neither fully natural nor artificial, neither historical reconstruction nor futuristic speculation. Instead, it behaves like a digitally generated fragment of an imagined history, engaging with symbolic, ornamental, and spatial richness. Its hyper-detailed surfaces recall geological strata and cultural sedimentation (Fig. 4.17), suggesting scalar complexity while pointing to restoration as a future practice shaped by algorithmic memory and computational craft.

Similarly, the *Fabricated Combines* project (Figs. 4.18 and 4.19) explores how digital fabrication and procedural design—a rule-based method of generating form through algorithms—can produce layered architectural assemblies that address both loss and continuity. The procedural logic of *Fabricated Combines* is shaped by image-based inputs that recall robotic archaeology and speculative ornament assemblages (Fig. 4.20), generating digitally fabricated artifacts assembled from fragments that evoke ruined materials and romanticized machinery. Through this synthesis, fractured elements are reassembled into new forms that register past use, cultural reference, and speculative projection, proposing not a return to the past but a re-invention; a materially situated, digitally mediated practice of historical engagement.

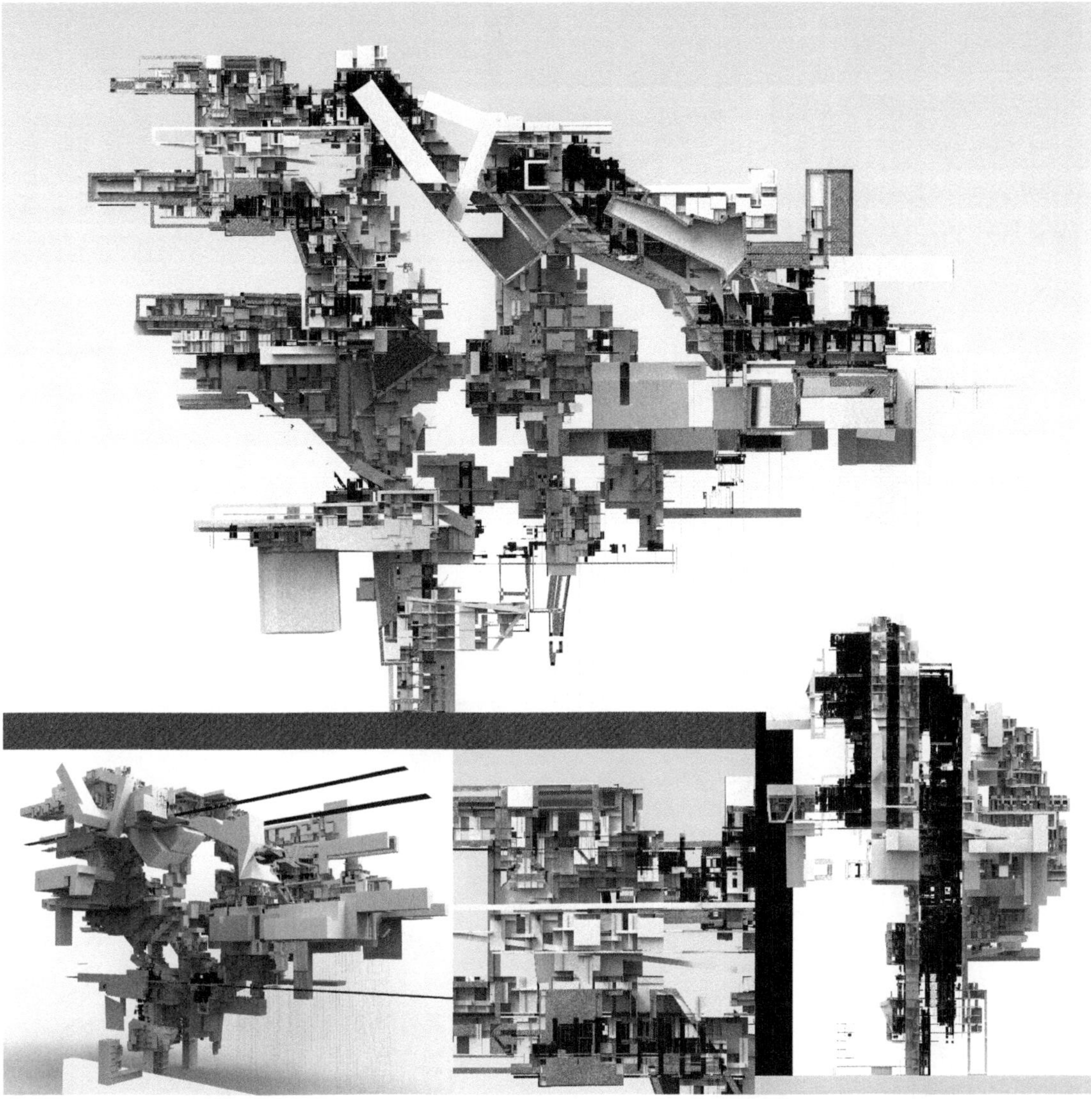

Figure 4.18 *Fabricated Combines*, digital drawings, 2025. Project by Sara Codarin and Masataka Yoshikawa.

Growth, Decay, Renewal

Buildings exist and persist in time. They are continuously exposed to the forces of nature—sun, wind, moisture, and biological life, and in turn, they are marked by those encounters. The natural environment acts upon the outer surface of architecture, leaving behind traces that become part

Figure 4.19 *Fabricated Combines*, 2025, 3D-printed model on display at the Venice Architecture Biennale, Arsenale, in the section on artificial creative practices. Project by Sara Codarin and Masataka Yoshikawa.

Figure 4.20 *Robotic Archaeology*, 2025, visual exploration exhibited at Pratt Institute's School of Architecture as part of the collective exhibition *Transduction: Artificial Intelligence in Architectural Experimentation*, curated by Jason Vigneri-Beane, Olivia Vien, Stephen Slaughter, and Hart Marlow. This series of speculative renderings contributed to the development of the *Fabricated Combines* workflow, alongside other visual and procedural inputs. Photo by Sara Codarin.

of the building's evolving identity. In this sense, architecture is never static, and its matter is shaped by erosion, aging, and transformation, reminding us that permanence is an illusion (Mostafavi and Leatherbarrow 1993). Weathering is traditionally viewed as a subtractive force. But can it also be understood as a generative agent? Can decay be additive? Can erosion enhance rather than erase? Throughout architectural history, elements like patina and rustication have been accepted—and justifiably romanticized—for their expressive value. Patina, as a kind of skin formed through layered accumulations of dust, moisture, microbes, and light, registers time, labor, and material endurance on a building's surface. Weathering is transformed into a visible record of ongoing interaction between architecture and environment. A recent installation titled

Stonelife[6], 2025, by Andrés Jaque and Gökçe Ustunisik, challenges the notion of preservation as defense against nature, and, instead, proposes the cultivation of new ecological relationships with it. Using digitally programmed vaporings, controlled bursts of water vapor, the installation activates microbial ecosystems on massive stone chunks, as metaphors for reframing buildings as living entities that breathe, evolve, and repair themselves.

Contributing to this ecological discourse, Barry Wark frames the built environment as a dynamic, evolving material condition and invites speculation about an ethic of restoration and whether it softens the nature-culture divide. Preservation, in this view, celebrates ongoing ecological processes rather than attempting to halt change (Fig. 4.21). His project *Nàdarra*7, 2024, exemplifies this sensibility through a full-scale

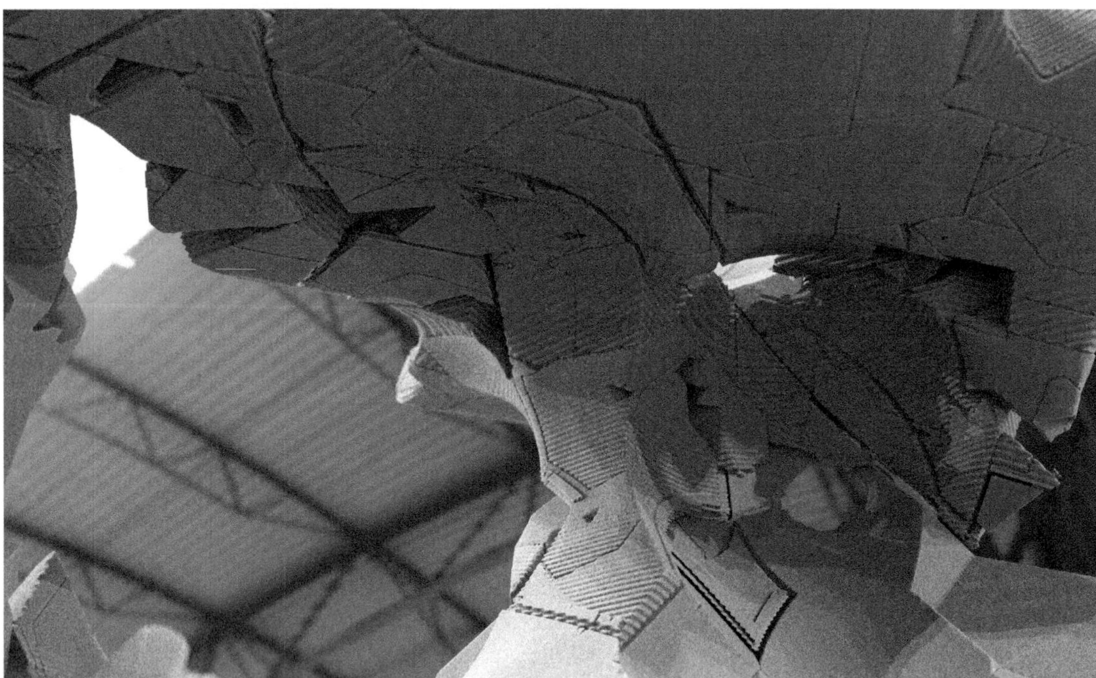

Figure 4.21 *Printed Parts.* 3D printed sand parts form a modular pavilion with an aesthetic of fragment, demonstrating the ability to be reappropriated or recycled into new elements, advancing ecological architecture through reuse, optimization, and adaptive reconfiguration. Courtesy of Barry Wark.

wall system made with 3D-printed sand and recyclable circular materials, designed to weather over time and support moss growth and ecological interaction (Fig. 4.22). The prototype draws on biophilic principles—multiscalar detail and organized complexity—challenging distinctions

Figure 4.22 *Nàdarra* installation. Composition inspired by ruins, grottos, and ancient architecture. Courtesy of Barry Wark.

between the natural and artificial, and proposing architecture as part of an evolving environmental system (Fig. 4.23). Here, ornament emerges not as a superficial overlay but instead as the result of simulated environmental forces, modeled through generative design tools. Ornament and weathering operate in tandem, forming a feedback loop through which ecological design and surface knowledge evolve gradually. The use of digital fabrication further blurs the boundary between the natural and synthetic, evoking processes such as stratification, erosion, and mineral growth and also engaging climate, decay, entropy, and regeneration as active design forces. The project gestures toward a broader aesthetic sensibility in which restoration acknowledges that weathering, aging, and decay are integral to architectural continuity.

Algae-Cellunoi[8], 2013, similarly fuses digital design, ecological systems, and ornament (Fig. 4.24). The installation is an ornamental wall structure composed of milled foam modules, described by the designers as a quintessential insulation material. What is typically hidden within external walls is turned inside out and revealed as a thick, ornate surface populated by terrestrial algae. This texture supports biological colonization, while liquid algae housed in 3D-printed flasks promote weathering and growth. Ornament is redefined as an environmental interface, a living archive that resists the utilization of predictable patterns and static formal conventions. By highlighting transformation and biological interaction, the project gestures toward an alternative understanding of preservation as an ongoing realignment with environmental processes. In this vision, the end of a building is never final, only a momentary phase in an unfolding material narrative shaped by nature, technology, and human care.

History Recoded

The rapid integration of artificial intelligence into the intellectual space of creative practices, combined with the growing availability of automated production tools, introduces new questions about how cultural artifacts are interpreted and reimagined. This shift points to the need for new forms of visual literacy that expand our understanding of historical art, design,

Figure 4.23 *Nàdarra* project. Detail showing the integration of architecture and constructed artifact. Courtesy of Barry Wark.

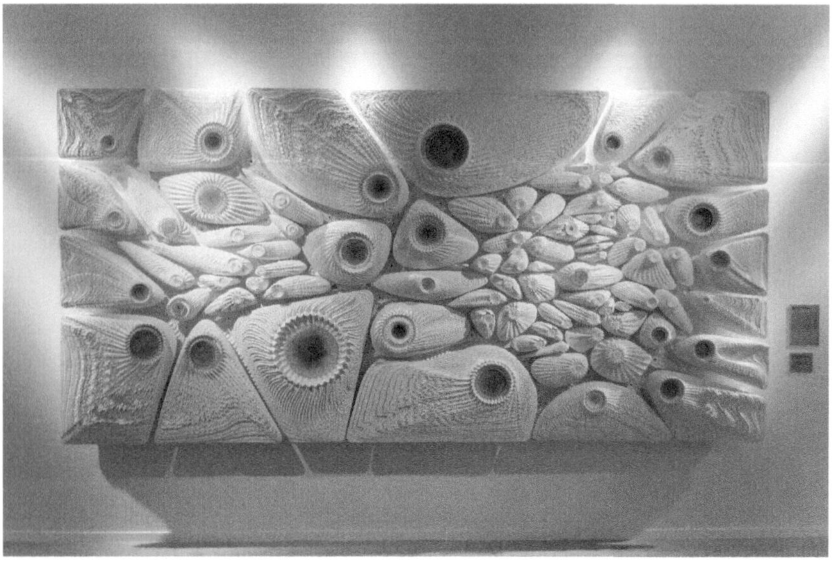

Figure 4.24 *Algae-Cellunoi* installation, produced for and exhibited at *9th ARCHILAB—Naturalizing Architecture*, FRAC Centre, Orleans France. Project by Marcos Cruz, Marjan Colletti, Guan Lee, and Richard Beckett. Courtesy of Marjan Colletti.

and architecture. Across history, visual styles have developed within specific cultural and political frameworks, embedding formal logics within their compositions. AI and machine learning offer tools to recognize and interpret these structures through computational analysis. As a result, relationships between past works and contemporary production can be read with fresh clarity, suggesting how historical forms continue to evolve aesthetic and cultural directions (Manovich 2018). This results in a newly expanded context, craft, tacit knowledge, and paradigm shift in design that may find new expression through computational codes. This development brings forward ongoing reflection on design and research strategies that incorporate digital technologies, algorithms, and advanced fabrication methods, while still maintaining an investment in human-centered thinking. Such tools, when applied critically, can contribute to a shared agenda rooted in creative, cultural depth. While predictive and technologically driven processes often operate through abstraction, their significance depends on being shaped by human engagement. The physical presence of designers—that carry knowledge, decisions, and embodied experiences grounds these workflows in material and cultural realities, consisting of context-specific approaches to creative work. Merging culture and technology through practice calls for design methods shaped by an epistemological perspective[9] and oriented toward focused, site-specific projects. At the same time, this approach intersects with the histories of augmented humanities, suggesting ways digital systems might be recoded to support collective intelligence.

The Serlio Code project brings some of these questions into material form through the prototyping of speculative artifacts. Through this process, AI is explored as a tool for interpreting design intentions that might otherwise remain inaccessible. By analyzing the architectural drawings of Italian Mannerist Sebastiano Serlio, the project applies machine learning to detect underlying formal structures and compositional strategies. These historical drawings are not treated as static documents but are reframed as generative systems capable of informing contemporary design speculation, giving them a rare gift of communication not available in the past. Computation now becomes a means of reengaging architectural history, surfacing latent patterns, and repositioning the archive within present-day discourse (Figs. 4.25 and 4.26).

Figure 4.25 *The Serlio Code*, 2022, Wedge Gallery, Woodbury University, Burbank, CA. Project by Jean Jaminet, Gabriel Esquivel, and Shane Bugni. Photograph by Zelig Fok. Courtesy of Jean Jaminet.

Figure 4.26 *The Serlio Code*, 2022. Texture detail of the artifacts. Photograph by Zelig Fok. Courtesy of Jean Jaminet.

Gothic Reimagined continues a parallel exploration through direct engagement with material processes. The project begins with 2D visuals created using the generative AI platform Midjourney, drawing on forms and details rooted in the Gothic architectural tradition. AI image generation is approached as a practice of working with digital content as training data, functioning as a kind of digital spolia[10] that reuses fragments of visual culture to generate new compositions (White 2024). These images are interpolated using depth maps, computational entities that convert pixel data into three-dimensional coordinates for model generation. The resulting digital geometry is then used as input for 3D printing to produce a stone sculpture, merging algorithmic invention with the material language of Gothic architecture. The outcome brings AI-generated imagery into dialogue with architectural craft, showing how digital processes can connect speculative design with historical material expression for the reuse and preservation of memory (Figs. 4.27 and 4.28).

Experimentation in the physical realm suggests that computational methods can reactivate historical visual logic, not through replication but

Figure 4.27 *Gothic Reimagined*, 2022, created by Dustin White (design lead) and Christopher Pope (design team), implemented with Concr3de 3D printing with powdered stone. Courtesy of Dustin White.

Figure 4.28 *Gothic Reimagined*, 2022. Additive manufacturing process. Courtesy of Dustin White.

through a generative and speculative lens in which history becomes an active site for design inquiry. Encounters between code, memory, and material practice trace ongoing and emerging possibilities for how the past might inform the evolving language of design.

Digitally Fabricated Replicas

Restoration faces a fundamental limitation: something that no longer exists cannot be brought back. When significant portions of an artifact's materiality are lost, any act of repair becomes an act of reconstruction. When full restoration or rehabilitation of historic material is no longer possible and reconstruction is pursued, the question often arises as to how, or whether, the lost elements should be reintroduced. In such cases, the creation of replicas becomes a recurring strategy that is considered acceptable when it extends beyond the goals of material conservation. In the Italian and broader European traditions, this shift from restoration to reconstruction has long been acknowledged. Reconstruction is approached with caution and is generally reserved for situations in which missing elements can be reintroduced without obscuring the historical stratification of the artifact.

By contrast, other international practices interpret restoration more expansively as the return of a building to a previous visual state.[11] This often entails reintegration and physical repair, with a focus on seamless continuity between old and new. New additions are designed to blend into an existing fabric, and the term "faithful" is frequently used to describe reconstruction efforts that prioritize visual coherence. Such continuity is commonly achieved through the fabrication of replicas—ornaments, architectural details, or even entire components—that help restore the perceived unity of the original work.

Italian restoration theory, instead, adopts a more nuanced stance in such contexts—defined as *conservative restoration*—advocating for a discernible approach in which any additions remain clearly distinguishable from the original. Replication is not excluded outright but rather reframed as a practice that must avoid imitative ambiguity, preserving both material and temporal legibility in a transparent manner.

So, in this view, replicas are not dismissed, but approached with greater care and intention, used as tools to engage critically with the traces of material loss. Both geometric and material replicas can support restoration practices in a variety of meaningful ways. As illustrated in the examples below, replicas can function as political gestures, support the dissemination of cultural knowledge, and contribute to the reactivation of spaces through conservative restoration, all while helping to recover the visual integrity, coherence, and symbolic presence of a lost artifact.

The Arch of Palmyra, reconstructed by the workshop Torart (Fig. 4.29) after the damage it suffered during the Syrian conflict in 2015, stands as

Figure 4.29 Robotic milling of a marble sculpture at the Torart workshop, where advanced automation meets traditional stonecraft. Courtesy of Torart.

an example of a politically charged replica. The structure was recreated through photogrammetry and 3D modeling based on digitized images. Since the site was inaccessible, the entire process occurred remotely, demonstrating how distributed knowledge and digital tools can support responses to cultural loss in times of crisis.

In 2004, at the archeological site of Pompeii, the replica of a wall covered in ancient graffiti was produced to support both research and public engagement.[12] *Graffiti in Pompeii* provided rare insights into the everyday life of a lost civilization originally found on fragile, non-durable surfaces. A full-scale digital reproduction was made using 3D scanning and modeling, then printed with gypsum powder. This version allowed for close study and interpretation of the Latin inscriptions without endangering the original. Details such as faint inscriptions and lighting effects became more accessible, revealing aspects that are often hard to perceive in situ. The replica served not only as a research object but also as a medium for storytelling, allowing the physical and virtual to work together in archaeological communication (Balzani et al. 2004).

An early example of digitally fabricated replication is the 2007 reproduction of the marble *Head of Maecenas* (Fig. 4.30) carried out by an Italian manufacturer in Pisa. A team of researchers[13] used laser triangulation to create a detailed 3D scan of the original, which was then processed into a digital model and physically carved using robotic milling on a marble block, followed by manual finishing to complete the fine details, including recreating the fractured nose by physically breaking it. The project, at

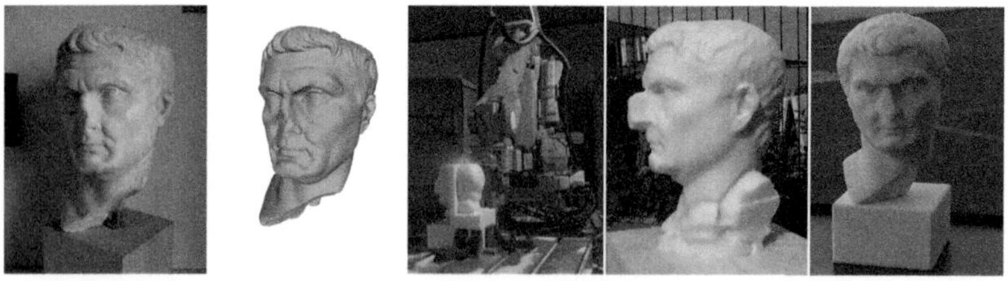

Figure 4.30 Reproduction process of the *Head of Maecenas* through subtractive robotic milling. The project was conducted by Institute of Information Science and Technologies (CNR-ISTI) and the SME Scienzia Machinale. Courtesy of the CNR-ISTI research group.

the time, aimed to test the precision of robotic fabrication and assess its potential for cultural use across a range of materials (Scopigno et al. 2017). The resulting replica was intended for scholarly use, allowing researchers to study the object while the original remained protected. This approach supports broader access to heritage without exposing fragile artifacts to the risks of transport or physical handling.

Quarra Stone, a company specialized in high-precision architectural stone fabrication, provides further examples of how replicas can blend digital techniques with craft traditions. The production facility works across different scales, from art installations to major architectural projects. One sculptural work, inspired by Michelangelo's Pietà, was created using concrete embedded with urban debris. The piece was robotically milled and then hand-finished to form a kind of contemporary fossil (Figs. 4.31 and 4.32). Quarra has also

Figure 4.31 Fabrication of Matthew Day Jackson's Pieta at Quarra Stone Company in Madison, Wisconsin. Courtesy of Quarra Stone.

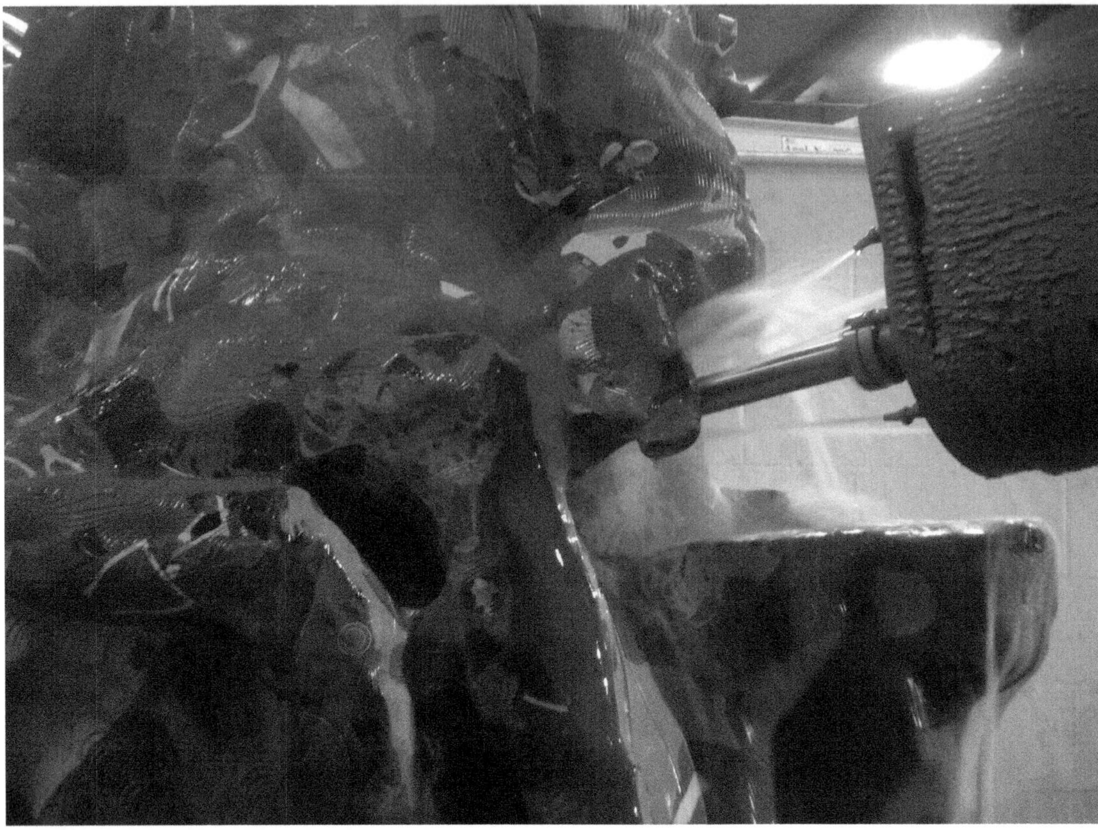

Figure 4.32 Close-up of Matthew Day Jackson's Pieta in fabrication at Quarra Stone Company in Madison, Wisconsin. Courtesy of Quarra Stone.

contributed to the restoration of buildings listed on the National Register of Historic Places, the official US inventory of historically significant sites. For the New York State Capitol, which was designed with a mix of French Renaissance Revival, Romanesque, and French Chateau styles, it reproduced ornate skylights and laylights in line with the original 1899 design. A similar approach was used for Harkness Tower in New Haven, an example of Collegiate Gothic architecture, where damaged stone elements were replaced with components fabricated using robotic milling and refined by hand. This combination of automated precision and manual craftsmanship preserved the architectural identity of these historic structures while also meeting current conservation standards.

Michigan Central Station in Detroit demonstrates how conservative restoration can include detailed reconstruction. After its closure in 1988, the building remained abandoned for decades and deteriorated extensively. It was added to the National Register of Historic Places in 1975, a designation that later helped prevent its demolition during years of neglect. A comprehensive restoration brought the structure back to working condition while respecting its original design.[14] Many of the station's ornamental elements had been damaged or lost over time. These features, part of its grand Beaux-Arts architecture and including the signature Guastavino tile vaults, were recreated through 3D scanning and printing. Thermoplastic polymer was used because it is lightweight and easy to install. Period lighting fixtures were also reproduced to match the historical appearance of the space (Fig. 4.33). The new parts are nearly indistinguishable from the originals. This approach reflects a restoration strategy that favors architectural continuity over visual contrast, which would have been more apparent had lower fidelity or visibly different materials been used. The renovated station now functions as a venue for exhibitions and public events and is set to host a mobility-focused innovation hub.

Digital Repair

Repair refers to the reconnection of fragments in order to restore the legibility of an artifact. The process is based on physical reconstruction of damaged or missing parts, starting from the identification and digitization of surviving fragments, which are then reassembled or complemented by newly fabricated elements. The objective is not only to regain the artifact's original outline of the volume but also to make its structure intelligible once again as a whole. Contemporary reconstruction increasingly relies on digital fabrication, both additive and subtractive. These modes of production stand as a counterpoint to anastylosis, which involves the reassembly of structures using original fragments. When those fragments are no longer available, reconstruction turns to new materials—ideally more durable and environmentally performative—and also to contemporary

Figure 4.33 The newly restored Michigan Central Station, 2024, with the restored Guastavino tile vaults in the Grand Hall. Photo by Sara Codarin.

construction techniques. In these cases, form is not recovered through the preservation of historical substance but through digital reproduction, which manifests as a fabricated, digitally driven form of "synthetic anastylosis."

A specific and symbolic instance of this approach—less concerned with repair than with continuation through the reconstruction of new elements—is the ongoing construction of the *Sagrada Família* church in Barcelona, originally designed by Antoni Gaudí. The building has advanced over the decades in alignment with Gaudí's vision of an artifact brought to completion across generations, a notion rooted in medieval building practices, where construction was understood as a collective effort unfolding over time, leaving entire communities to feel invested in the realization of a monument. Gaudí acted as a master builder, relying not on conventional architectural drawings but on notational information embedded in physical models to communicate design intent. The project has adapted to shifting construction techniques and craft traditions, integrating robotic technologies and CAD/CAM translations to fabricate architectural components based on Gaudí's original physical models (Burry 2012).

At a smaller scale, an example of digital repair is the restoration of the *Madonna di Pietranico*, a severely damaged terracotta sculpture whose volumetric integrity was compromised by the earthquake that struck Abruzzo, Italy, in 2009. In this project, researchers[15] explored additive manufacturing as a viable solution for reconstructing the artifact. The aim was to fabricate precise components capable of reconnecting the surviving fragments. Because the fracture surfaces were eroded and key components were missing, traditional recombination through gluing was not feasible. The workflow combined data collection, 3D modeling, and 3D printing, beginning with the digital scanning of the broken pieces and the processing of spatial data necessary to determine their relative positioning (Fig. 4.34). Additive manufacturing, efficient for producing custom connectors with complex geometries and irregular interfaces, allowed for the creation of a lightweight support structure that could integrate seamlessly with the original material. Comparisons with historical photographs provided an additional method of validation, ensuring the accuracy of the support volume and informing multiple assembly tests. During the modeling phase, the team accounted for the irregular shapes of the surviving fragments while developing a non-invasive and reversible intervention strategy. A 3D printer equipped with nozzles was used to deposit thin layers of gypsum and synthetic binder, forming a support

Figure 4.34 Digital reconstruction of the *Madonna di Pietranico*. Courtesy of the CNR-ISTI research group.

structure compatible with terracotta. As per restoration best practices, the intervention remains detectable, the geometry is intentionally distinguishable, and the action is fully reversible. The process ultimately restored the legibility of the sculpture in its overall image, reestablishing the *unità potenziale*—the potential unity, or oneness—of the work.

Looking forward, developments such as multi-material printing and the customization of interface geometries will further refine the relationship between original and added matter, improving the legibility of repaired volumes and expanding conservation strategies. The significance of these developments is confirmed by the European Union's support of research into advanced repair methods. Among them, the *RePAIR* project *(Reconstructing the Past: Artificial Intelligence and Robotics meet Cultural Heritage)* aims to integrate AI and robotics into archaeological reconstruction. Since many archaeological artifacts survive only as fragmented remains, *RePAIR* seeks to streamline their physical reassembly by automating what is typically a frustrating process of manual labor. Through automation, the project aims to open new pathways for recovering damaged cultural heritage at scale, enhancing both efficiency and precision in restoration efforts.

The Role of the Architect: New Skills for the Digital Master Builder

Repair, reconstruction, and restoration are all oriented toward conservation, each establishing a relationship with the material traces of the past and inevitably generating transformations that are both material and figurative (Dalla Negra 2017). When discussing the new technologies available for intervening on historical matter, a common concern arises: the risk of technocratic application, which can be seen when tools begin to dictate design decisions or when default procedures override the specific needs of the artifact. Historically, there have been cases where the identity of a contemporary design has overshadowed the historical value of the historic object itself, like the iconic restoration of Castelvecchio Museum by the architect Carlo Scarpa. Manfredo Tafuri, in this regard, provocatively assigns a transformative role to restoration as an architectural act. He even claims to "tolerate" the interventions of Carlo Scarpa who, despite "aggressively altering" monuments, was capable of producing works of great significance "by the grace of God," but this was the result of a rare talent and should not be mistaken for a general license to pursue creative freedom at the expense of historical substance (Tafuri 1991). The danger arises when architects believe they possess this grace: every means becomes justified in the name of design authorship, potentially opening the door to subjectivity and a technocratic drift.[16] For this reason, a distinction must be made between signature restorations—authored interventions, or *restauro d'autore*—and standard professional practice. The latter should remain grounded in rigorous, methodological approaches, applicable across scales: from the object, to the building, and into the urban fabric.

History has shown that our global dependence on technology, and the momentum of technological advancement, cannot be reversed. Contemporary culture is defined by the fluid transfer of digital innovation across industries, where knowledge diffusion and blurred disciplinary boundaries are the norm. While architecture is rooted in long-standing traditions, the definition of the architectural professional must now confront evolving technologies, interconnected tools, and shifting modes of production. The challenge lies in finding space for critical authorship while adapting skillsets that span across digital and physical realms.

For centuries, an architect's expertise has required a deep understanding of construction. As far back as the Middle Ages, the designer-craftsperson was responsible for managing all aspects of building, from formal composition to the development of structural systems. This proto-professional figure was known as the *master builder*, evolving from an even earlier role, that of the *master mason*. The master builder embodied the roles of architect, builder, engineer, and scientist, using accumulated experience to train future generations of craftspeople. In turn, these successors refined tools and techniques, generating innovation that was grounded in the knowledge of prior histories. One of the most significant master builders in architectural history was Filippo Brunelleschi, who

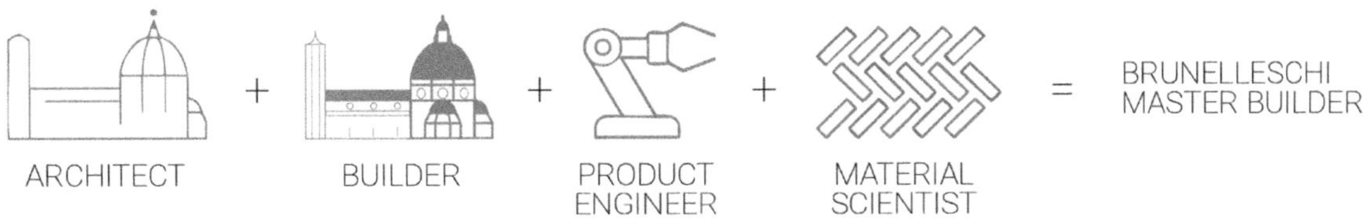

Figure 4.35 The relative simplicity of building technologies during the Renaissance allowed Filippo Brunelleschi to operate as a master builder, overseeing design, engineering, and construction within a unified role.

oversaw the construction of the dome of Santa Maria del Fiore in Florence (Fig. 4.35). He advanced the use of traditional materials through technical ingenuity, for example, by developing a herringbone brick-laying pattern and reinforcing the drum with iron chains. These interventions defined not only the dome's structural system but also the technologies necessary for its construction, including machinery for lifting materials (King 2013; Di Pasquale 2002).

Architectural design and production begin to diverge during the Renaissance, "when designers began to make drawings" (Mitchell and McCullough 1995). This shift coincided with the invention of perspective and orthogonal projections as tools for representing architecture. The act of drawing redefined the conception-production relationship, positioning the architect as an intellectual authority. Drawings became instructions—abstracted directives from the thinker to the maker—marking a transition in which ideas were elevated above craft. This separation between design and built work widened during the Industrial Revolution with the emergence of the contractor and the professional engineer. The rise of serial production marginalized the designer's creative involvement, eliminating the need for material knowledge and direct engagement with fabrication and isolating the roles from one another. Centralized, experiential knowledge gave way to standardized and automated processes. Replication, defined by the production of standardized copies, replaced imitation as the dominant mode of making. In the 20th century, increasing complexity in building systems and the proliferation of new materials led to a growing specialization of roles. Architectural authorship became diffused across a constellation of consultants. The architect was redefined as a *master-controller,* one who "coordinates the many diverse consultants who are able to master their own specialties" (Kieran and Timberlake 2004). The adoption of digital workflows in professional practice has allowed architects to use computational data to inform building production. Design time previously dedicated to drawing has been restructured, as digital models now closely approximate built reality.

In the current phase, what has been described as the Fourth Industrial Revolution, conception and production are once again reunited. Following the phases of hand-making, mechanical-making, and digital-making, the integration of digital fabrication technologies has allowed architects to "almost instantaneously produce scale models of their design using processes and techniques identical to those used in the industry" (Kolarevic 2004). Unlike previous industrial paradigms that prioritized standardization, tools such as robotics and CNC fabrication enable customization and variation. This technological alignment between design and production reactivates the figure of the *master builder*, reimagined as one who develops a material sensitivity that belongs to the manual skill of ancient craftspeople through the medium of the machine (Figliola 2017). This evolving figure can be identified as a *digital artisan-architect* or an *information master builder*, capable of generating construction data directly from design information. In this framework, the architect becomes a *digital master builder*, one who can bypass redundant steps in the design-to-execution pipeline and reclaim a central role in the act of making.

The relationship between designer and machine has revived early visions of mechanization in architecture, such as Louis Kahn's notion of construction equipment, like cranes, functioning as extensions of the architect's body[17] (Kahn 2003). In this vision, mechanical systems begin to substitute for human musculature, translating intention into material action with programmable precision through flexible, adaptable mechanisms (Scheurer 2012). Among these, the robot occupies a distinct position, not just as a tool for automation, but as a machine that reintroduces a form of material engagement through a mediated yet direct process.

Robots are not limited to executing predefined instructions; they also have the potential to reshape how architects conceptualize space, structure, and fabrication. They promote a spatial intelligence that moves beyond the constraints of traditional drawing and frontal projection, encouraging design thinking that is multi-directional and responsive to forces like gravity. With their kinematic flexibility, robots enable operations that are directly tied to design intent and material behavior. This shift allows architects to move beyond the confines of the drawing sheet and return to thinking at a 1:1 scale, where form is developed through direct engagement with process and material. As a result, artifacts can be materialized through the action of "powerful artificial arms" (Picon 2014), operating at the intersection of digital and physical, process and outcome, computational control and sensory awareness. Robotic fabrication supports a more iterative, tactile, and process-oriented approach, where architectural outcomes emerge through material behavior, procedural intelligence, and real-time interaction. This shift expands the ability to work with complex material conditions and, at the same time, redefines the role of the designer as one who must adapt, think procedurally, and make decisions in response to evolving processes.

This renewed material intelligence enabled by robotics becomes particularly relevant when considered within the context of architectural restoration, where design must respond not only to abstract parameters but to the physical constraints, precision, and craftsmanship demanded by the site. The building restoration site presents a highly specific context in which intervention typically requires detailed on-site labor, intensive effort, precision, and assembly tailored to existing conditions. Due to the uniqueness of these interventions, it is often difficult to incorporate standard construction components or rely on typical building practices. Architectural heritage restoration remains strongly tied to craft traditions, with deep knowledge held by skilled artisans. While essential, these traditions can disrupt digital continuity in the transition from design to realization. To bridge digital processes and craft-based practices in contemporary restoration, architects need to operate across multiple domains, drawing on the knowledge of design, construction, engineering, and material science. This calls for the operations of a *digital master builder*, a role that has the potential to unify the fragmented aspects of architectural production by combining intellectual, technical, and manual modes of working into a cohesive practice, "in a single mind, heart, and hand" (Kieran and Timberlake 2004). Developing this role further requires the pursuit of an integrated design methodology, where creative and technical processes are understood as parts of a continuous workflow, linking design, material knowledge, and fabrication logic. An additional opportunity emerges when considering the architect as an *algorithm builder* (Burger 2012). In this expanded role, the architect does more than define form; they also create rule-based systems that generate both geometry and fabrication logic. These algorithms manage the transition between digital design and physical realization by incorporating complex parameters, allowing for an evolving exchange in which inputs and outputs are defined as part of the same process.

This newly conceived architectural practice brings renewed attention to the role of craftsmanship. As Richard Sennett observes, craftsmanship is not defined by the act of physically manipulating materials, but by intrinsic motivation and a commitment to doing a job well for its own sake, paired with a cultivated "material consciousness" (Sennett 2008). Even as digital tools become increasingly adaptable and sophisticated, there remains a critical place for individual interpretation and authorship. Architects are still responsible for determining how best to engage these tools within specific design and construction contexts. The renewed focus on material engagement also brings relevance to David Pye's concept of "workmanship," which he defines as any process in which the quality of the result is not predetermined, but instead relies on the maker's judgment, dexterity, and care throughout the act of making (Pye 1968). This perspective resonates with the ongoing development of a fabrication-aware design culture, in which digital tools function not as neutral intermediaries but as active participants that shape form and material behavior. The foundational skills of architectural practice are not being lost; they are instead being extended and refined through interaction with digital systems, creating new ways to connect conceptual design with embodied making and creating new inspiration along the way.

Notes

1 In 1964, during the Second International Congress of Architects and Specialists of Historic Buildings, thirteen resolutions were adopted. The first established the Venice Charter, and the second led to the founding of the International Council on Monuments and Sites (ICOMOS).

2 Minimum intervention (1931 Athens Charter, 1932 Italian Charter, 1964 Venice Charter, 1972 Italian Charter, 2000 Krakow Charter, 2003 ICOMOS Charter); Reversibility (1972 Italian Charter, 2000 Cracow Charter, 2003 ICOMOS Charter); Compatibility of the integration with the pre-existence (1972 Italian Charter, 2000 Krakow Charter, 2003 ICOMOS Charter); Recognizability of the new intervention (1931 Athens Charter, 1972 Italian Charter, 1964 Venice Charter, 1972 Italian Charter, 2003 ICOMOS Charter); Readability of the formal unity (1931 Athens Charter, 1972 Italian Charter, 2003 ICOMOS Charter); Case-by-case approach (1931 Athens Charter, 1932 Italian Charter, 1972 Italian Charter, 2003 ICOMOS Charter); Use of modern materials and construction techniques (1972 Italian Charter, 2003 ICOMOS Charter); Digital documentation (2003 ICOMOS Charter; 2019 Declaration of Cooperation on advancing digitization of Cultural Heritage).

3 In the context of cultural heritage or architecture, an ontology helps describe not just what elements are present in a 3D model (like walls, columns, or frescoes), but also how they relate to each other, their historical context, materials, function, and transformations over time. For example, it might define that a "column" is a type of "structural element," often made of "stone," and part of a "classical order."

4 In architecture, incrustation refers to a decorative technique in which a contrasting material is applied to a surface, either as an inlay or an overlay, to create visual richness and texture.

5 The term *phenomenological world* refers to the lived experience of space as perceived through the body and senses, with attention to atmosphere, material presence, and the ways architecture shapes human perception.

6 The installation Stonelife was showcased in the Natural Intelligence section of the 2025 Venice Architecture Biennale, curated by Carlo Ratti.

7 Nàdarra is a Scots Gaelic word meaning "natural," used to describe something inborn, inherent, or originating from nature.

8 Algae-Cellunoi is a compound term combining "algae" (from Latin alga, meaning seaweed) with "cellunoi," referencing cellular structures and Voronoi patterns.

9 Epistemological perspective refers to the underlying assumptions about the nature, origin, and scope of knowledge that inform a particular method, framework, or approach. In design and creative practice, it shapes how information is interpreted, what is considered valid knowledge, and how decisions are made based on that understanding.

10 Spolia refers to the reuse of architectural elements, materials, or decorative fragments from earlier buildings in new constructions, often carrying symbolic, aesthetic, or practical significance. The term originates from the Latin *spolium*, meaning "spoils," and historically implies both appropriation and transformation.

11 Terms like preservation, conservation, renovation, rehabilitation, and restoration are often used interchangeably, though each technically and conceptually carries distinct implications. For example, what is commonly referred to as *restoration* in international contexts, typically meaning the, is known in the Italian tradition as *conservative restoration*, a definition that implies a more specific scope within a wider set of practices.

12 The research was conducted by DIAPReM and CNR-ISTI.

13 The German Federal Ministry of Education and Research commissioned the reproduction of the head of Maecenas to the CNR—Institute of Information Science and Technologies in Pisa and the Scienzia Machinale Group, as part of the research program "Maecenas."

14 The project, completed in June 2024, was led by Quinn Evans Architects in collaboration with Ford Motor Company, which owns the property. The project brought together a multidisciplinary team of architects, engineers, historians, and conservation specialists to rehabilitate the monumental architecture.

15 The restoration project was completed under the coordination of the Superintendence for Historical, Artistic, Artistic and Ethno-anthropological Heritage of L'Aquila, with the collaboration of CNR-ISTI Visual Computing Lab (Pisa), Tryeco 2.0 s.r.l. (Ferrara), Equilibrarte s.r.l. (Rome).

16 It is worth noting that, particularly within the Italian context, the official stance of the *Soprintendenza* (heritage authority) is one of strict conservative restoration. While historical layering is now celebrated as a rich architectural and cultural palimpsest, similar interventions would be considered unacceptable if carried out today. Iconic examples such as the Basilica of San Clemente in Rome, with its three superimposed churches, the Cathedral of Syracuse, built atop a Doric temple, and Piazza Navona, constructed over the foundations of the Stadium of Domitian, reflect historical processes of transformation that would not align with current preservation standards.

17 In his 1061 "Essays," Louis Kahn notes: "one day I visited the site during the erection of the prefabricated frame of the building [...] Now I'm glad of this experience because it made me aware of the meaning of the crane in design, for it is merely the extension of the arm like a hammer. Now I began to think of members 100 tons in weight lifted by bigger cranes."

References

Abbott, Alison. 2017. "The 'Time Machine' Reconstructing Ancient Venice's Social Networks." *Nature* 546 (7658): 341–44. https://doi.org/10.1038/546341a.

Balzani, Marcello, Marco Callieri, Matteo Fabbri, et al. 2004. "Digital Representation and Multimodal Presentation of Archeological Graffiti at Pompei." *VAST* 4: 93–103.

Bonsma, Peter, Iveta Bonsma, Silvia Parenti, and Pedro Martín Lerones. 2016. "Inception Standard for Heritage BIM Models." *Lecture Notes in Computer Science* (Cham), 590–99. https://doi.org/10.1007/978-3-319-48496-9_47.

Brandi, Cesare. 1963. *Teoria del Restauro*. Edited by Licia Vlad Borrelli and Joselita Raspi Serra. Ed. di storia e letteratura.

Burger, Shane M. 2012. "Algorithmic Workflows in Associative Modeling." *Digital Workflows in Architecture,* Edited by Scoot Marble, 132–48.

Burry, Mark. 2012. "Models, Prototypes and Archetypes Fresh Dilemmas Emerging from the'file to Factory'era." In *Manufacturing the Bespoke: Making and Prototyping Architecture* Edited by Bob Sheil. John Wiley and Sons.

Carbonara, Giovanni. 1997. *Avvicinamento al Restauro: Teoria, Storia, Monumenti*. 1st ed. italiana. Liguori.

Carbonara, Giovanni. 2012. "An Italian Contribution to Architectural Restoration." *Frontiers of Architectural Research* 1 (1): 2–9. https://doi.org/10.1016/j.foar.2012.02.007.

Carpo, Mario. 2017. *The Second Digital Turn: Design Beyond Intelligence*. With Cynthia Davidson. Writing Architecture Series. The MIT Press.

Cayer, Aaron. 2018. "From Archive to Office: The Role of History in Theories of Architecture Practice." *Ardeth* 2 (March): 35. https://doi.org/10.17454/ardeth02.04.

Chatterjee, Anuradha. 2019. *John Ruskin and the Fabric of Architecture*. Routledge.

Codarin, Sara, and Karl Daubmann. 2022. "VR Gestural Modeling to Recapture the Human Body in Design." In 110th ACSA Annual Meeting Proceedings: Empower, 458–65. https://doi.org/10.35483/acsa.am.110.63.

Colletti, Marjan. 2010. "Exuberance and Digital Virtuosity." *Architectural Design* 80 (2): 8–15. https://doi.org/10.1002/ad.1036.

Colletti, Marjan. 2017. "Artist's Essay: Post-Digital Neo-Baroque: Reinterpreting Baroque Reality and Beauty in Contemporary Architectural Design." In *Neo-Baroques*. Brill.

Colletti, Marjan. 2024. "Postdigital Neobaroque. TRANSMEDIA PRACTICES – Volume 2." Universität Innsbruck. https://doi.org/10.25651/1.2022.0008.

Colletti, Marjan, and Peter Massin. 2024. "Postdigital Neobaroque. FRAGMENTARY ACTS OF TRANSFERENCE – Volume 1." Preprint, Universität Innsbruck. https://doi.org/10.25651/1.2022.0007.

Colletti, Marjan, Peter Massin, Laura Winterberg, and Viktoria Hörtenhuber. 2024. "Postdigital Neobaroque. TRAITS OF POSTDIGITAL NEOBAROQUE – Volume 3." Universität Innsbruck. https://doi.org/10.25651/1.2022.0009.

Dal Co, Francesco. 2013. "Scienziati Del Restauro e Architetti Felici." *Casabella*, no. 18.

Dalla Costa, Mario, ed. 2005. *Memoria e Restauro dell'Architettura: Saggi in Onore di Salvatore Boscarino*. Ex fabrica 4, [Testimonianze]. FrancoAngeli.

Dalla Negra, Riccardo. 2017. "Architettura e Preesistenza: Quale Centralità?" In *Architettura e Preesistenze. Premio Internazionale Domus Restauro e Conservazione Fassa Bortolo*. Skira.

Daubmann, Karl. 2021. "The Role of the Hand in Digital Design." In *Digital Fabrication in Interior Design*. Routledge.

Daubmann, Karl, and Lauren Jacobi. 2019. "From String to Volume." In *Finding San Carlino*. Routledge.

Di Giulio, R., F. Maietti, E. Piaia, M. Medici, F. Ferrari, and B. Turillazzi. 2017. "Integrated Data Capturing Requirements for 3D Semantic Modelling of Cultural Heritage: The Inception Protocol." *The International Archives of the Photogrammetry, Remote Sensing and Spatial Information Sciences* XLII-2/W3 (February): 251–57. https://doi.org/10.5194/isprs-archives-xlii-2-w3-251-2017.

Di Pasquale, Salvatore. 2002. *Brunelleschi: La Costruzione Della Cupola Di Santa Maria Del Fiore*. Marsilio.

Di Stefano, Elisabetta. 2010. "Ornamento e Architettura. L'estetica Funzionalistica Di Louis H. Sullivan." *AESTHETICA*. 89 (89): 7–60.

Figliola, Angelo. 2017. "Post-Industrial Robotics: Exploring Informed Architectures in the Post-Digital Era." *TECHNE – Journal of Technology for Architecture and Environment*, August 4, 256–66. https://doi.org/10.13128/TECHNE-19751.

Floridi, Luciano. 2015. "Hyperhistory and the Philosophy of Information Policies." *The Onlife Manifesto*, 51.

Germanà, Maria Luisa. 2014. "Technology and Architectural Heritage. Research Experiences in Archaeological Sites." *TECHNE – Journal of Technology for Architecture and Environment*, May 21, 41–51. https://doi.org/10.13128/TECHNE-14531.

Germanà, Maria Luisa. 2019. "Technology and Architectural Heritage: Dynamic Connections." In *Conservation of Architectural Heritage*, Edited by Dean Hawkes, Hocine Bougdah, Federica Rosso, et al. Advances in Science, Technology & Innovation. Springer International Publishing. https://doi.org/10.1007/978-3-030-10871-7_7.

Kahn, Louis. 2003. "Form and Design." *Louis Kahn, Essential Texts*, 64. Robert Twombly, W. W. Norton Company Ltd.

Karch, Tristan, Jakhongir Saydaliev, Isabella Di Lenardo, and Frédéric Kaplan. 2025. "LLM-Powered Agents for Navigating Venice's Historical Cadastre." Version 1. Preprint, arXiv. https://doi.org/10.48550/ARXIV.2505.17148.

Kieran, Stephen, and James Timberlake. 2004. *Refabricating Architecture: How Manufacturing Methodologies Are Poised to Transform Building Construction*. McGraw-Hill.

King, Ross. 2013. *Brunelleschi's Dome: How a Renaissance Genius Reinvented Architecture*. Bloomsbury Publishing USA.

Kolarevic, Branko. 2004. *Architecture in the Digital Age: Design and Manufacturing*. Taylor & Francis.

Körner, Andreas. 2024. "Postdigital Incrustations and Procedural Sinters." In *Postdigital Neobaroque. TRANSMEDIA PRACTICES*, Edited by Marjan Colletti, vol. 2. Universität Innsbruck. https://resolver.obvsg.at/urn:nbn:at:at-ubi:3-40900.

Lee, Wei-Han Vivian, and James Macgillivray. 2023. "Assemblies of Labor." In *104th ACSA Annual Meeting Proceedings: Shaping New Knowledges*, Edited by Robert Corser and Sharon Haar. Association of Collegiate Schools of Architecture.

Lenardo, Isabella di, and Frédéric Kaplan. 2015. "Venice Time Machine: Recreating the Density of the Past." https://infoscience.epfl.ch/handle/20.500.14299/122077.

Logothetis, S., A. Delinasiou, and E. Stylianidis. 2015. "Building Information Modelling for Cultural Heritage: A Review." *ISPRS Annals of the Photogrammetry, Remote Sensing and Spatial Information Sciences* II-5/W3 (August): 177–83. https://doi.org/10.5194/isprsannals-ii-5-w3-177-2015.

Lorusso, Salvatore. 2002. *La Diagnostica per Il Controllo Del Sistema Manufatto-Ambiente (Diagnostics for the Control of the Manufactured-Environment System)*. Pitagora Editrice.

Lorusso, Salvatore, Giovanni Carbonara, and Maria Teresa Gentile. 2002. "Il Restauro Architettonico: le Diverse Concezioni nel Corso dei Secoli." *Quaderni di Scienza della Conservazione* 2: 7–18.

Lynn, Greg. 2004. "The Structure of Ornament." In *Digital Tectonics*, Edited by Neil Leach, David Turnbull, and Chris Williams. Wiley–Academy.

Maietti, Federica, Roberto Di Giulio, Marcello Balzani, Emanuele Piaia, Marco Medici, and Federico Ferrari. 2017. "Digital Memory and Integrated Data Capturing: Innovations for an Inclusive Cultural Heritage in Europe through 3D Semantic Modelling." In *Mixed Reality and Gamification for Cultural Heritage*. Springer.

Maietti, Federica, Roberto Di Giulio, Marcello Balzani, Emanuele Piaia, Marco Medici, and Federico Ferrari. 2018. "3D Data Acquisition and Modelling of Complex Heritage Buildings." In *Lecture Notes in Computer Science*. Springer International Publishing. https://doi.org/10.1007/978-3-319-75826-8_1.

Maietti, Federica, Roberto Di Giulio, Emanuele Piaia, Marco Medici, and Federico Ferrari. 2018. "Enhancing Heritage Fruition through 3D Semantic Modelling and Digital Tools: The INCEPTION Project." *IOP Conference Series: Materials Science and Engineering* 364 (June): 012089. https://doi.org/10.1088/1757-899x/364/1/012089.

Maietti, Federica, Marco Medici, Federico Ferrari, Anna Elisabetta Ziri, and Peter Bonsma. 2018. "Digital Cultural Heritage: Semantic Enrichment and Modelling in BIM Environment." In *Digital Cultural Heritage*. Springer.

Manovich, Lev. 2018. *AI Aesthetics*. Strelka Press Moscow.

Mitchell, William J. 1998. *The Logic of Architecture: Design, Computation and Cognition*. 6. print. MIT Press.

Mitchell, William J., and Malcolm McCullough. 1995. *Digital Design Media*. John Wiley & Sons.

Morton, Timothy. 2013. *Hyperobjects: Philosophy and Ecology after the End of the World*. University of Minnesota Press.

Mostafavi, Mohsen, and David Leatherbarrow. 1993. *On Weathering: The Life of Buildings in Time*. MIT Press.

Nofal, Eslam, Rabee M. Reffat, and Andrew Vande Moere. 2017. "Phygital Heritage: An Approach For Heritage Communication." Preprint, Verlag der Technischen Universität Graz. https://doi.org/10.3217/978-3-85125-530-0-36.

Otero-Pailos, Jorge, Erik Langdalen, and Thordis Arrhenius, eds. 2016. *Experimental Preservation*. Lars Müller Publishers.

Pallasmaa, Juhani. 2009. *The Thinking Hand: Existential and Embodied Wisdom in Architecture*. AD Primers. Wiley.
Picon, Antoine. 2014. "Reinventing the Meaning of Ornament." In *Ornament*, 1st ed. Wiley. https://doi.org/10.1002/9781118658321.ch4.
Picon, Antoine. 2014. "Robots and Architecture: Experiments, Fiction, Epistemology." *Architectural Design* 84 (3): 54–59. https://doi.org/10.1002/ad.1754.
Pye, David. 1968. *The Nature and Art of Workmanship*. Vol. 4. Cambridge University Press.
Saunders, Andrew. 2009. "Baroque Parameters." *Architectural Design* 79 (1): 132–35.
Saunders, Andrew. 2017. "Baroque Topologies: Novel Approaches to Analysis and Representation of the Baroque Interior in the Era of Big Data." In *INTBAU International Annual Event*. Springer.
Saunders, Andrew. 2018. *Baroque Topologies*. Palombi editori.
Scheurer, Fabian. 2012. "Digital Craftsmanship: From Thinking to Modeling to Building." *Digital Workflows in Architecture* 76.
Scopigno, R., P. Cignoni, N. Pietroni, M. Callieri, and M. Dellepiane. 2017. "Digital Fabrication Techniques for Cultural Heritage: A Survey." *Computer Graphics Forum* 36 (1): 6–21. https://doi.org/10.1111/cgf.12781.
Sennett, Richard. 2008. *The Craftsman*. Yale University Press.
Sposito, Alberto, ed. 2004. "La Conservazione Affidabile per il Patrimonio Architettonico. Reliable Conservation of Architectural Heritage." Flaccovio.
Stellacci, Stefania, Danilo Giglitto, and Chiara Piccoli. 2024. *Advanced Research and Design Tools for Architectural Heritage: Unforeseen Paths*. Taylor & Francis.
Tafuri, Manfredo. 1991. "Storia, Conservazione e Restauro: Una Intervista a Cura Di Chiara Baglione e Bruno Pedretti." *Casabella : Rivista Mensile Di Archite Ura e Tecnica*, no. 580: 23–26.
White, Dustin. 2024. "Digital Spoliare." *Diffusions in Architecture: Artificial Intelligence and Image Generators*, 180–90.

Chapter 5

Disembodied Tools for Restoration

DOI: 10.4324/9781003590873-5

Case Studies

This section seeks to bring together two research areas: architectural restoration and robotic fabrication. These fields have much to offer one another, yet they often operate within distinct disciplinary frameworks and rely on different expertise. The approach begins with a comparative analysis of best-practice case studies in architectural restoration, focusing on projects that required material reconstruction or the addition of new elements after loss, damage, or disassembly. The projects extend beyond preservation or conservation and embrace contemporary interventions that allow for the emergence of new aesthetic languages while still maintaining respect for historical fabric. In parallel, this section examines practices in robotic fabrication, reflected as both a set of tools for custom production and as platforms for digital data capture. These programmable, general-purpose machines offer a level of precision that can approach or expand upon traditional craft. The selected projects highlight technical possibilities and limitations of each method. Placing cultural heritage and robotics case studies side by side reveals potential points of convergence (Fig. 5.1). The goal is to identify where methodologies might inform one another and to suggest new hybrid approaches. At present, fully automated construction remains largely experimental and is far from being a widespread, regulated practice. Its application in restoration, a field that demands precision, historical sensitivity, and case-by-case decision-making, is even further from becoming viable. Restoration, as Cesare Brandi reminds us, concerns "special products," fundamentally different from "industrial products" (Brandi 1963). Restoration is guided by cultural and historical values, not just functional requirements. Recovering structural stability or usability may form part of the intervention, but the primary objective will always be to preserve the meaning and identity embedded in the built work. Therefore, rather than propose a fully robotic future for restoration, the aim is to explore how the principles of conservation might intersect with emerging technologies. Robotic systems should be understood as tools that extend the knowledge and skills of craftspeople

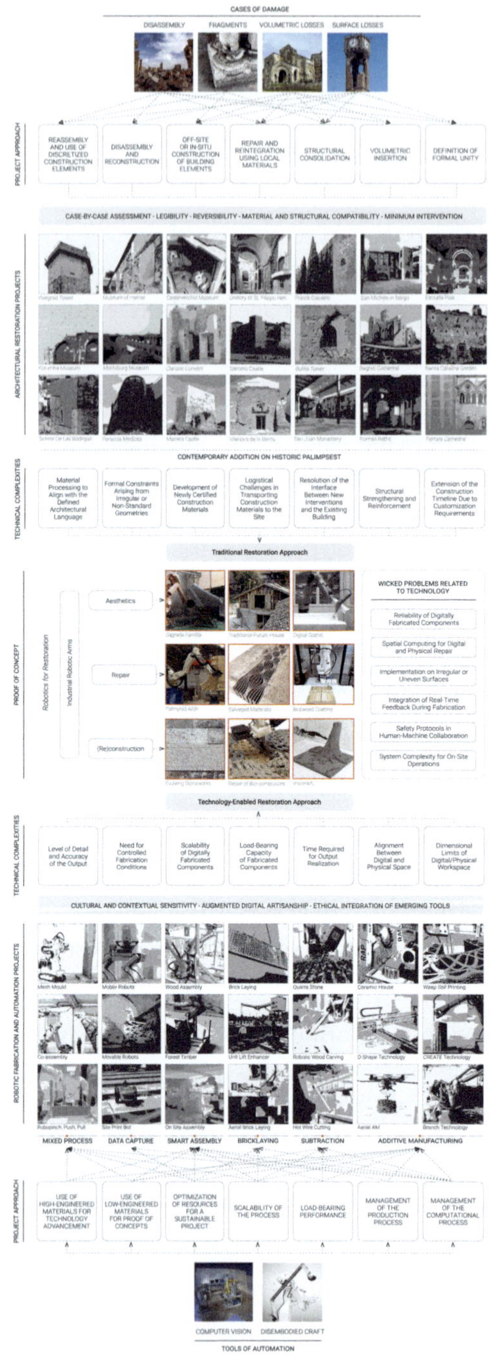

Figure 5.1 Atlas of analyzed examples in architectural restoration and robotic fabrication. The diagram presents two parallel fields: restoration practices grounded in historical continuity and cultural significance, and robotic fabrication techniques developed for custom, automated production. Their convergence is explored through selected case studies that apply robotic arms to experimental construction, restoration, or speculative reinterpretations of historical forms and craft-oriented techniques. These examples illustrate how digital tools can engage with material heritage through informed, culturally situated processes of making.

and conservation experts. These technologies can support sensitive site-specific tasks, assist in making informed material decisions, and contribute to the continued transmission of cultural heritage by expanding opportunities and design possibilities.

Architectural Restoration

Project Selection and Analysis Criteria

The selected case studies focus on the relationship between temporal layers of the built environment, where the historical artifact is understood as a complete and unrepeatable work of human cultural production—unrepeatable precisely because it belongs to the past—and must be critically positioned in relation to the present (Minguzzi 2008; Balzani 2011; Balzani and Dalla Negra 2017). In exploring this relationship, Giovanni Carbonara, Italian professor and international reference in architectural conservation, proposes a framework for classifying practical approaches to architectural intervention on existing structures, particularly in the insertion of contemporary elements within historical contexts. The framework is structured around three modes of intervention: autonomy and dissonance, assimilation and consonance, and a dialectical relationship that seeks the reintegration of the architectural image (Carbonara 2011).

The first of these, autonomy and dissonance, reflects a deliberate separation between old and new. In this approach, the contemporary addition asserts its autonomous independence, often through contrasting forms or materials, in order to avoid falsifying the historical narrative. Assimilation and consonance, seek to harmonize the new with the old, integrating contemporary elements in a way that produces formal and material unity. Between these two positions lies the dialectical approach (Fig. 5.2), which recognizes both the autonomy of contemporary design and the continuity of the historical structure. This mode of intervention does not rely on imitation or opposition, but instead opens a critical dialogue that reintegrates the image of the building while allowing for creative interpretation. Each intervention approach requires a clear distinction

Figure 5.2 Oratory of San Filippo Neri, Bologna, Italy. Restoration project by Pier Luigi Cervellati and Giorgio Volpe, 1999–2000. The intervention reconstructed the original barrel dome through a contemporary timber rib structure, integrating mid-century repairs with new structural reinforcements to convert the space into an auditorium. Photo by Sara Codarin.

between the conservation phase, which focuses on preserving the historical material of the artifact, and the design phase, which involves the informed and contemporary transformation of the existing architecture. When the intervention consists of adding matter without any subtraction from the historical fabric, the restoration becomes a "legitimate transformation" (Carbonara 1997), grounded in both respect and innovation.

The case studies illustrate how contemporary architectural language can be effectively used to integrate technological components through material and formal transformations that serve conservation purposes. Just as a traditional design process is grounded in critical inquiry and expressed through creative acts, the restoration process similarly manifests through careful, case-specific gestures. These may involve the preservation of material in the state it exists today, or the functional reinterpretation of preexisting elements—at times through the reintegration of the architectural image and, where appropriate, through the use of contemporary design language (Dalla Negra 2011). This analysis takes an intellectual position on architectural expression, advocating for the intentional use of innovative production. Contemporary language can be meaningfully connected to the existing elements of a building through an approach based on reinterpretation rather than replication, and resists making new elements appear old, instead placing them confidently within a contemporary context (De Angelis d'Ossat 1995). In moments when restoration directly addresses the reintegration of a missing part, a clear dialogue emerges between traditional craft and present-day technologies.

The architectural restoration projects are analyzed in response to the specific types of damage affecting architectural elements, and they are compared in terms of the scale of the contemporary addition. They range from individual technological units that can be localized, self-contained components like custom infill elements or discrete repair parts, to technological subsystems, which involve coordinated assemblies that serve broader functions. These subsystems can be prefabricated façade panels, composite floor structures, or even full architectural systems where the intervention engages with the building's primary structural or spatial organization like roofing, circulation, or enclosure systems (Zaffagnini 1981). Further case study observations consider the materials employed and the design approach adopted and are intentionally evaluated through the same critical lens—guided by autonomy and dissonance, assimilation and consonance, or a dialectical relationship oriented toward reintegrating the architectural image (Fig. 5.3).

This framework for presenting the case studies made obvious the operational complexities inherent in restoration processes, including geometric constraints, logistical challenges involved in fabricating specialized building components for restoration and the compatibility of new materials with existing building elements. Other factors include performance assessments, level of detail and color treatment of the new volumetric insertions, logistical requirements for transporting materials to the restoration site, and the need for highly specialized labor or on-site procedures during assembly and installation phases. At large, site activities can range in technical complexity. Simpler interventions require only basic tools and are usually managed locally, often as preventive maintenance tailored to individual sites. More complex interventions demand specialized design knowledge, advanced techniques, and the use of sophisticated equipment. These situations may benefit from centralized planning across multiple sites, making larger-scale maintenance efforts more efficient and cost-effective (Germanà and Sposito 2001). These technical demands can be supported through the use of innovative tools that offer new possibilities for improving precision, coordination, digital data storage, and long-term performance in restoration practice.

Robotic Fabrication

Project Selection and Analysis Criteria

The case studies in robotic fabrication examine a variety of automated systems, primarily involving industrial robotic arms, as a means to demonstrate the technical capacities of digitally mediated making (Brell-Cokcan and Braumann 2013; McGee and Ponce De Leon 2014; Reinhardt et al. 2016;

Figure 5.3 Diagram of restoration case studies categorized by type of damage, intervention scale, material strategy, and design approach. Projects range from localized components to full architectural systems. Each is assessed for fabrication complexity, site logistics, and cultural integration, following a framework of autonomy, assimilation, or dialectical reintegration within historical contexts.

ARCHITECTURAL RESTORATION

Project	Designer(s)	Location	Year	Damage				Scale			Component					Primary material									Approach			Intervention				Primary technical complexities								
																dry						wet																		
				volumetric loss	wall gap	damaged fragment	disassembled tectonics	technological unit	technological subsystem	architectural system	exterior wall	interior wall	floor	roof	independent unit	stone	brick	wood	glass	metal	ceramic	concrete	mortar	clay	autonomy/dissonance	assimilation/consonance	dialectical reintegration	continuous/monolithic	fragmented/discretized	load-bearing	non load-bearing	geometric constraints	customization	structural enhancement	material compatibility	high-fidelty detailing	aesthetic language	transportation efforts	implementation logistics	on-site operations
1 Visegrad Tower	János Sedlmayr	Visegrad, Hungary	1964	•	•			•	•		•	•		•				•		•		•					•		•	•	•	•	•			•	•		•	•
2 Museum of Hamar	Sverre Fehn	Hamar, Norway	1973	•	•			•	•		•	•	•	•		•			•			•			•				•	•	•	•	•		•					•
3 Castelvecchio Museum	Carlo Scarpa	Verona, Italy	1973	•	•	•		•	•		•	•	•	•		•		•	•	•		•					•		•	•	•	•	•	•	•	•	•			
4 San Filippo Neri Oratory	Pier Luigi Cervellati, Giorgio Volpe	Bologna, Italy	2000	•	•	•	•	•	•		•	•	•	•				•									•		•	•		•	•	•			•		•	•
5 Medieval Cassero	Riccardo Dalla Negra, Pietro Ruschi	Prato, Italy	2000			•			•		•							•		•							•		•	•				•	•	•	•			
6 Convent of San Michele in Borgo	Massimo Carmassi	Pisa, Italy	2001	•		•				•	•	•	•	•	•		•					•			•				•	•	•	•		•	•		•			
7 Escuela Pias de San Fernando	Linazasoro & Sánchez	Madrid, Spain	2004	•	•				•					•				•		•		•					•		•	•		•		•	•	•				•
8 Kolumba Museum	Peter Zumthor	Cologne, Germany	2007	•		•				•	•	•	•	•		•	•		•	•			•		•				•	•		•	•	•	•	•	•		•	•
9 Moritzburg Museum Extension	Nieto Sobejano Arquitectos	Halle, Germany	2008	•		•				•	•	•	•	•	•				•	•		•			•			•		•		•	•		•		•			•
10 St. Anthony Former Church	2TR Architettura	Santa Fiora, Italy	2009		•		•		•	•	•	•	•	•	•	•							•				•	•		•		•	•	•	•		•	•	•	•
11 Saliceto Castle	Armellino and Poggio Architetti	Saliceto, Italy	2011	•	•				•	•					•			•	•	•					•				•	•		•			•					•
12 Bofilla Tower	Fernando Vegas, Camilla Mileto	Valencia, Italy	2011	•	•	•			•		•		•					•		•				•			•			•	•	•		•	•	•	•			•
13 Bagrati Cathedral	Andrea Bruno	Bagrati, Georgia	2012	•	•	•	•			•	•	•	•	•					•	•					•				•	•		•	•	•	•			•	•	•
14 Santa Catalina Botanical Garden	Isuuru Architects	Sierra de Badaya, Spain	2013	•	•	•	•	•			•							•									•		•	•	•	•	•				•		•	•
15 Setenil De Las Bodegas Tower	Fernando Visedo	Setenil De Las Bodegas, Spain	2014	•	•			•	•		•	•	•	•		•						•				•			•	•	•	•			•		•			•
16 Fortezza Medicea	De Vita and Schulze Architects	Arezzo, Italy	2015	•	•				•		•							•							•				•	•	•	•	•					•	•	•
17 Matrera Castle	Carquero Arquitectura	Villamartin, Spain	2015	•	•	•	•	•	•		•					•						•	•				•			•	•	•			•		•			
18 Vilanova de la Barca Church	AleaOlea Architecture & Landscape	Vilanova de la Barca, Spain	2016	•	•		•		•		•	•	•	•			•				•		•		•				•	•		•		•	•	•	•			•
19 Monastery of San Juan	BSA	Burgos, Spain	2017	•	•	•	•	•	•				•	•				•							•			•		•		•		•			•	•	•	•
20 Sao Pedro do Sul Roman Baths	Joan Mendes Ribeiro	Sao Pedro do Sul, Portugal	2019	•		•		•	•		•		•	•		•		•									•	•		•		•		•			•		•	•
21 Ferrara Cathedral	Studio Berlucchi	Ferrara, Italy	(2035)			•		•	•		•			•		•										•			•		•	•	•	•	•	•		•		•

Willmann et al. 2019). Some of these tools are already in use on restoration sites for surveying and diagnostic purposes, while others present opportunities for bespoke production tailored to the needs of damaged historical architecture. Robotic digital fabrication establishes a direct connection between parametric design workflows and material execution, enabling the development of context-specific methodologies. Across these examples, the productivity of automated systems is tested to support precision, responsiveness, and continuity between analysis and intervention.

The case studies are organized into two macro-categories of automation: disembodied craft and computer vision. Disembodied craft refers to robotic systems that perform tasks based on a predefined digital design input, executed in a static and unidirectional manner. These systems separate the act of design and the moment of fabrication both temporally and geographically.[1] The designer's gestures or intentions are encoded digitally and later reproduced by the robot, often in a different place or at a different time (Figs. 5.4 and 5.5). The robot does not interpret or adapt; it simply follows instructions and functions as mechanical extension of the designer's arm and workspace. The setup is generic and not dependent on specific toolpaths, end-effectors, or kinematics. From an engineering perspective, these "non-intelligent robots" are the most straightforward: a fixed sequence of operations is repeated without variation, and no sensors are involved to provide feedback or enable correction.

In contrast, computer vision systems enable robots to respond to their environment and make decisions during execution. These systems use sensing technologies, such as 2D and 3D scanners, to collect spatial and visual data in real time. This allows the robot to interpret its surroundings, recognize forms, and continuously compare actual output to the intended design. As a result, the toolpath can be adjusted dynamically, introducing a feedback loop that supports a more adaptive and interactive mode of fabrication. These "higher-intelligence robots" move beyond static execution and participate actively in the making process. This model can parallel aspects of skilled human labor, for instance, how a bricklayer adjusts each gesture in response to the changing consistency and weight

Figure 5.4 An example of a disembodied craft-making exercise is the *Robo-Pinch, Push, Pull* project, 2014, at the Taubman College of Architecture and Urban Planning, University of Michigan. A robotic arm was used to mediate the dexterity of material behavior through plaster casting. Courtesy of Karl Daubmann.

of mortar, guided by the ability to incorporate visual and tactile feedback. Contemporary robotic systems can now learn from such human behaviors, memorizing and reproducing movement sequences that incorporate variation.[2] This introduces the possibility of a new kind of robotic craft capable of precision combined with responsiveness.

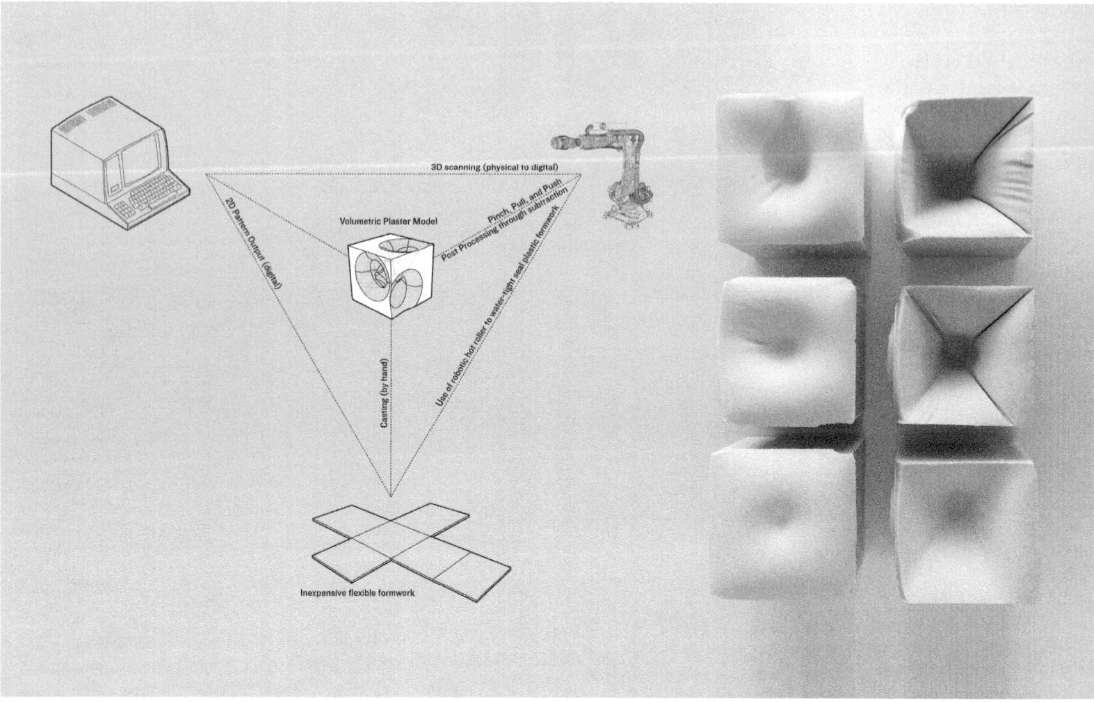

Figure 5.5 *Robo-Pinch, Push, Pull* project, 2014. Parametric variation achieved through iterative plaster casting. Courtesy of Karl Daubmann.

The selected projects represent a range of robotic approaches to architectural production, including additive manufacturing, subtractive manufacturing, smart assembly, combined processes, bricklaying, and data capture (Fig. 5.6). Additive manufacturing refers to processes that translate digital information into physical form, producing complex morphologies with structural optimization potential. These systems typically rely on powder-bed deposition or extrusion and accommodate a broad range of materials. Subtractive manufacturing, defines the final outcome through the removal of material from a solid volume, commonly using spindles, hot-wire cutters, or chisels on materials such as wood, stone, or ceramic. Smart assembly involves the use of robotic arms, typically equipped with grippers, for the installation or precise placement of components. These systems are frequently used for constructing masonry

Figure 5.6 Diagram of robotic fabrication case studies involving disembodied craft and computer vision systems. The projects reflect a range of approaches including additive and subtractive manufacturing, smart assembly, bricklaying, and data capture. Each is evaluated for material strategy, scale of application, and system maturity within digitally-mediated architectural production.

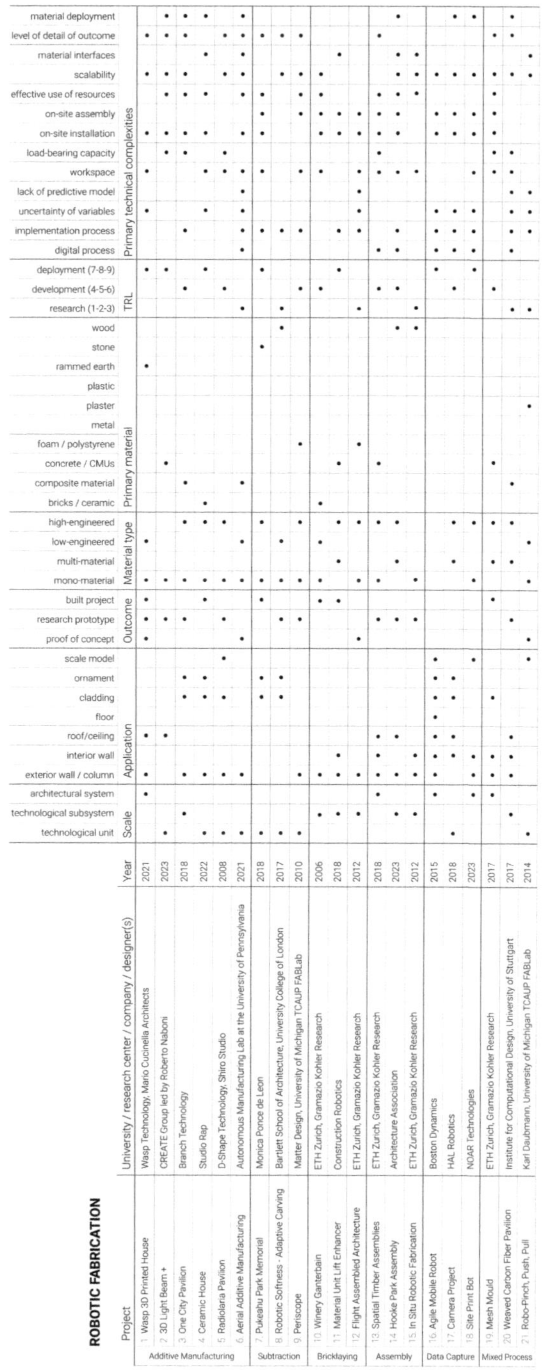

walls from brick or stone, as well as assembling dimensional lumber or steel. Combined processes integrate additive, subtractive, and assembly methods to expand the operational range of robotic systems. Bricklaying is a specialized method focused on the precise placement of discrete modular elements like bricks, into patterned assemblies that ensure structural performance and spatial articulation. Lastly, data capture includes the use of mobile robots and drones to collect environmental and geometric information, often feeding into sensing or adaptive fabrication loops.

Robotic-enabled projects in this study were also evaluated in terms of their scale of application, ranging from individual units to subsystems and full architectural assemblies, as well as their material strategies and technology readiness level (TRL).[3] This multi-scalar approach helped assess the maturity and applicability of each system, identifying process innovation (nonlinear workflows), tools (sensor-integrated and site-adaptable robotics), and design (form-generation through automation). Observations broadly revealed key technical complexities inherent in automation, including the orchestration of computational workflows, the consistency of performance outcomes, and the optimization of resources within the pursuit of material efficiency and environmental responsibility. These challenges are especially relevant in the context of restoration and bespoke construction, where robotic systems can be leveraged to reduce human error while maintaining a high degree of dexterity and control over form. Ultimately, these examples point toward the potential of fabrication-aware thinking in heritage contexts, where workflows extend from digital decision-making to in situ material production. As robotic systems begin to augment and extend human sensory capacity, the long-term vision becomes clear: to reimagine traditional craftsmanship as a digitally mediated practice capable of operating directly on-site, harnessing its strengths rather than replacing them.

Robotics for Restoration

The convergence of projects in architectural restoration and digital fabrication made it possible to identify a subset of nine relevant works that engage with historical references in diverse and meaningful ways (Fig. 5.7).

	Project	Scale	Material	Type of Intervention	Robotic Process
(Re)construction	1. Sagrada Familia	[L]	Stone	Digital Continuity	Robotic Stone Carving
	2. Arch of Palmyra	[M]	Stone	Symbolic Reconstruction	Robotic Milling
	3. Evolving Stoneworks	[S]	Concrete, Reclaimed Aggregates	Disaster Recovery	Concrete Extrusion
Repair	4. Traditional House of the Future	[L]	Concrete	Vernacular Revitalization	On-Site Robotic Construction
	5. Salvaged Materials	[M]	Biobased Composites	Material Reuse	Hydrogel Extrusion
	6. Repair of Biocomposite Materials	[S]	Biobased Composites	Material Repair	Conformal Robotic Printing
Aesthetic Language	7. Digital Gothic	[M]	Foam, Plaster	Form Finding Ornament	Robotic Hot Knife Cutting
	8. Biobased Coating	[M]	Biobased Composited	Bio-Based Coating	Robotic Hydrogel Deposition
	9. Viscera/L	[M]	Clay	Behavioral Fabrication	Clay Toolpath Shaping

Figure 5.7 Overview diagram of nine case studies illustrating the intersection of robotics and restoration. Projects are evaluated by scale of intervention and categorized by type, (re)construction, repair, and aesthetic language.

These examples highlight the use of industrial robotic arms for fabrication, applied across a range of materials and construction techniques. Each project supports the hypothesis that restoration can be reimagined through high-tech processes, not as a break from tradition, but as a means of extending it. The selected examples demonstrate a shared commitment to technologically advanced fabrication approached in a conscientious way, guided by methodologies rooted in human culture and attentive to the continuity of material traditions. Their relevance to cultural heritage is not limited to site or form, but is embedded in an ethos of making that questions how contemporary tools operate within historically significant contexts, and where robotic construction can be a promising and conceptually charged methodology. Industrial robots are considered here as mechanical extensions of the human arm, capable of precision, consistency, and dexterity. Their adaptation to restoration practice signals a larger disciplinary shift: the reinterpretation of tools developed for industrial automation toward an architecture grounded in humanity and values. The robot, a symbol of standardization and mass production, is recontextualized within one of the most delicate and site-specific domains of architectural labor. This intentional juxtaposition between automation and care points at the urgency of establishing robust methodologies to accommodate new technologies that enter the design space.

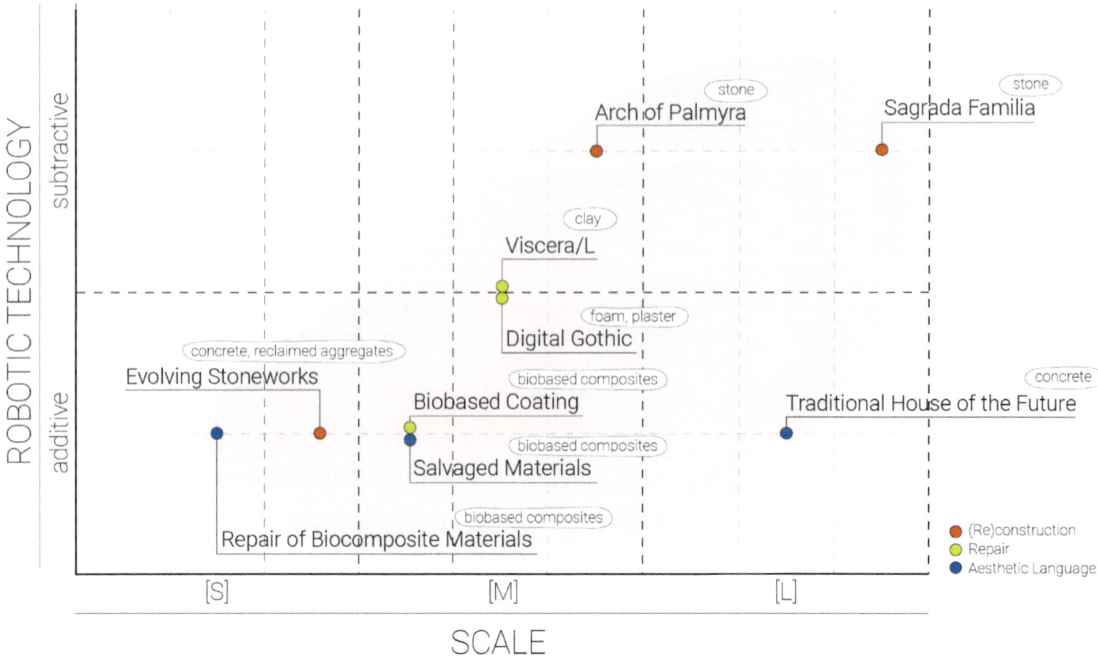

Figure 5.8 Diagram mapping nine case studies across the categories of (re)construction, repair, and aesthetic language, with each positioned by its scale of intervention and robotic process.

The *Robotics for Restoration* case studies [1–9] are organized into three categories: (re)construction, repair, and aesthetic language. Each example is also evaluated by its scale of intervention, identified as small, medium, or large based on the scope of the outcome (Fig. 5.8). The category of (re)construction includes projects that use robotic fabrication to continue ongoing construction, produce replicas of historical elements, or reuse demolition materials to reconstruct damaged or collapsed buildings. Projects classified under *repair* demonstrate how robotic construction can support both physical and symbolic acts of reparation. These examples explore how fragmented materials can be reconnected, how experimental materials can be stabilized or healed, and how architectural interventions can contribute to restoring damaged social or cultural fabrics through deliberate material and spatial strategies. Projects focusing on aesthetic language investigate how robotics can be used in an artisanal manner, allowing precise coding to coexist with material irregularity and intentional imperfection. These works explore form-finding processes that draw from historical patterns, geometries, and compositional logic, as a digitally informed continuity of architectural expression.

A quick overview of the curated project work includes: [1] The Sagrada Familia in Barcelona, where robotic stone carving continues Gaudí's vision through parametric modeling and digital tooling, enabling the completion of the famously unfinished complex geometries. [2] The reconstruction of the Arch of Palmyra in Carrara that demonstrates how robotic milling and 3D scanning can produce symbolic replicas in response to heritage loss. [3] The Italian startup *Centauroos* utilizes robotic arms to extrude material sourced from demolition waste, creating a closed-loop reconstruction process that addresses both material scarcity and disaster recovery. [4] The *Traditional House of the Future* uses on-site robotic fabrication to integrate concrete printing with vernacular timber construction in rural China, proposing a replicable model for rural renewal. [5] Chalmers University's reuse workflows combine robotic vision and 3D printing to adapt damaged timber, producing hybrid joints and surface reintegration that extends the life of salvaged components. [6] The Copenhagen-based research center CITA's conformal printing system enables the repair of weathered biocomposite panels, treating decay as a generative design parameter. [7] At MIT, hot-knife robotic carving reinterprets Gothic modes of making, producing cast elements that retain expressive toolpath traces. [8] Robotic hydrogel coatings on timber surfaces also developed at Chalmers offer a sustainable and culturally responsive method for surface repair, where irregularities emerge as design features. [9] In *Viscera/L*, implemented at the University pf Pennsylvania, additive and subtractive robotic processes shape clay through real-time sensing and feedback, producing variable textures that redefine ornament as an active, material dialogue.

Case Study 1 | (re)construction

Robots at the Sagrada Familia: Digital Gaudí

The ongoing construction of the Sagrada Familia Church in Barcelona exemplifies how computational design workflows and the robotic production of complex geometries have evolved over several decades, advancing the volumetric integration of an architectural pre-existence. The main structure is now approaching completion; the towers and core architectural elements are expected to be finalized by 2026, while decorative features and a large stairway leading to the main entrance are projected to continue until 2034. The continuation of Antoni Gaudí's project stands as a literal extra-ordinary example, unique in both scope and character, of contemporary advancement in the construction of an internationally recognized architectural icon. Few projects in the world match its scale, stylistic singularity, or the sheer complexity of its Gothic Revival building elements and their organic nature. The Sagrada Familia's ongoing construction demands the long-term coordination of human, technical, and economic resources in ways that are unparalleled elsewhere. The professionals engaged in the project exemplify process optimization and the application of advanced technologies, using computational tools to geometrically define and carve massive stone elements in a continued effort to complete a historical structure of exceptional cultural and architectural value.

The lead architect guiding the continuation of the Sagrada Familia is Mark Burry, a New Zealand-born designer and academic who has played a central role in the project since 1979. His involvement began as an architecture student working on his final thesis, during which he met two practitioners, former students of Gaudí, who were overseeing construction at the time. This encounter led to an enduring collaboration that has shaped the trajectory of the project ever since. Over more than four decades, Burry has dedicated himself to deciphering Gaudí's compositional strategies and directing the project through successive waves of technological development, moving from manual drawing methods to advanced computational modeling. Under his direction, key elements such as the principal Glory Façade and parts of the Passion Façade, including the rose window and the narthex, were designed and completed in 2017. He also oversaw the development of the Sala Creuer, an auditorium situated above the crossing where the six main towers converge with the body of the basilica.

Although Antoni Gaudí lived during the 19th and early 20th centuries and witnessed Barcelona's transformation into an industrial city, his architectural imagination was deeply rooted in medieval ideals of social and spatial organization. This perspective shaped his approach to the Sagrada Familia, where he envisioned not only a religious structure but a revival of the medieval tradition of building. Much like Filippo Brunelleschi's role in constructing the dome of Santa Maria del Fiore, Gaudí intended to supervise the work closely, acting as both designer and artisan, communicating his design intent directly to the builders through hands-on demonstrations rather than detailed drawings.

As a result, much of Gaudí's design intent was embedded in physical models rather than conventional architectural documentation. When Mark Burry joined the project, his initial role was to study and interpret these models, primarily made of plaster, many of which had been damaged during the Spanish Civil War. This work required reverse-engineering and extracting data from the surviving fragments to produce accurate drawings. With the advent of 3D modeling software, Burry transitioned from manual reconstruction to computational representation. In his 1996 paper *Parametric Design and the Sagrada Familia*, Mark Burry explains that the digital control of complex geometries—such as doubly ruled second-order surfaces, helicoids, hyperbolic paraboloids, and hyperboloids of revolution—required advanced computational techniques to ensure precision and feasibility in completing Gaudí's architectural project (Burry 1996). To address these challenges, high-end modeling software originally developed for the aerospace industry was adapted for architectural use, enabling representation and manipulation of relevant geometries (Fig. 5.9).

Figure 5.9 Computational design studies by Mark Burry interpreting Gaudí's geometric language for the Sagrada Família, exhibited as part of *The Analects of Gaudí* at the 2025 Venice Architecture Biennale. Photo by Sara Codarin.

In the chapter *Models, Prototypes and Archetypes: Fresh Dilemmas Emerging from the "File to Factory" Era*, Mark Burry discusses the digitization and fabrication of several architectural elements of the Sagrada Familia, including the nave columns, the rose window of the Passion Façade, and the Glory Façade (Burry 2012). These efforts required the development of a digital workflow focused on uncovering generative principles in Gaudí's design language, using intersecting polygons, convex and concave parabolas, and rotational patterns based on triangles, squares, and hexagons. Once these principles were understood and full-scale dimensions inferred from the physical models, the information was translated into precise digital definitions, enabling finalization for robotic fabrication and production at an architectural scale.

Work in the digital environment became a critical exercise of tectonic analysis and design evaluation, made systematic through the parametrization of geometric primitives. Since tectonics relies on structural and material choices to shape architectural expression, the use of computational tools allowed for the exploration of formal systems in ways that were not feasible in Gaudí's time. These tools overcame the geometric and material constraints that he could only approach intuitively. Digital methods increased the likelihood of realizing Gaudí's ambition of high customization, because many architectural elements, such as columns and windows, were designed to be unique in this project. This same logic extended into robotic fabrication, employing a circular saw for stone cutting as an actuator to streamline production, thereby controlling timelines and labor costs. The process materialized as a contemporary form of digitally assisted stereotomy, a craft tradition already present in the original construction of the Sagrada Familia, where stone blocks were shaped with precision to fit geometry-intensive elements (Burry 2016). Tasks that once demanded extensive hand carving and significant labor could now be carried out through computational modeling and machine-guided cutting. As a result, the project retained the expressive materiality of stone construction while reducing the manual effort and time traditionally required.

The construction of the Sagrada Familia exemplifies the transition from analog to digital workflows across decades, evolving from manual reconstructions to advanced computational modeling and robotic fabrication. These tools enabled large-scale prototyping that served not only as geometric testing but also as interpretive validation of Gaudí's formal language.[4] The physical models became critical instruments of discovery and verification, especially in historically complex projects where little conventional documentation remains. A significant technological leap occurred with the shift from 6-axis to 7-axis robotic arms, allowing for the machining of larger, more intricate stone components—sandstone, granite, basalt, and porphyry—while also minimizing the need for secondary assembly. Design decisions in this process were deeply shaped by fabrication constraints, shifting the designer's role toward that of a digital toolmaker. As Burry emphasizes, automation increases—rather than reduces—the designer's responsibility as file-to-factory workflows bypass traditional intermediaries. Although Gaudí originally maintained that innovation should reside in design, not in construction, he came to accept the need for new materials and methods late in life (Burry 2005).

Today, the Sagrada Familia stands as perhaps the most representative case of robotics-heritage integration: not a reconstruction or repair, but the continuation of an unfinished project through computational design and high-fidelity digital manufacturing. Beyond its symbolic and technical importance, the project also raises the possibility that digitally coordinated, craft-informed production could eventually extend beyond singular, high-profile landmarks and begin to influence design and construction at a broader infrastructural scale. As a culturally significant testbed, the intervention on the Sagrada Familia demonstrates how digital fabrication and traditional craftsmanship can be brought into dialogue, with the artisan remaining a vital collaborator within an increasingly automated environment. This post-digital condition does not replace craft but redefines it embedding human expertise within technologically mediated workflows, and presenting itself as a new and promising extension of artistry.

Case Study 2 | (re)construction

Reconstructing Memory: Robotic Fabrication of the Arch of Palmyra

The reconstruction of the Triumphal Arch of Palmyra aimed to produce a scaled replica of the original structure from the archaeological site in Syria. This replica was created in Carrara by carving stone blocks using a six-axis robotic milling system and later presented in London for exhibition. The individual pieces, including detailed decorative elements, were then assembled to recreate the full arch. This project serves as an example of replicating a culturally significant monument and was carried out in response to the destruction caused by the Syrian conflict in 2015.

Palmyra has been recognized as a UNESCO World Heritage Site since 1980 and was officially listed as endangered in 2013. The site contains the monumental remains of an ancient city that played a cultural role during the first and second centuries when it was part of the Roman province of Syria. One of its key structures, the Triumphal Arch of Palmyra, was an ornamental gateway built in the

3rd century with limestone during the reign of Emperor Septimius Severus. Palmyra gained importance as a city located on trade routes connecting Persia, India, and China to the Roman Empire, making it a meeting point for different civilizations. Its ruins were rediscovered by European travelers in the 17th and 18th centuries and its architecture combines Greco-Roman styles with local and Persian influences, especially visible in its funerary sculptures, created for tombs and burial monuments. UNESCO designated the site as a World Heritage Site for its exceptional artistic value and its role in demonstrating important cultural exchanges over time.

Since being placed on the List of World Heritage in Danger, UNESCO has monitored Palmyra to assess damage and understand the broader consequences of cultural loss. Threats to such sites often stem from natural disasters or human activity that disrupts the physical and cultural fabric of a place. In the case of Palmyra, the main threats have included destruction from armed conflict (Fig. 5.10), deliberate

Figure 5.10 The archaeological site of Palmyra, showing the remains of the Temple of Bel. The photograph was taken after conflict-related events caused damage to several Roman-era structures. Courtesy of the Directorate-General of Antiquities and Museums, Syria.

Figure 5.11 Damaged artifacts at the National Museum in Palmyra. Courtesy of the Directorate-General of Antiquities and Museums, Syria.

targeting of heritage (Fig. 5.11), illegal excavations, and a lack of funding for conservation work. These losses not only affect showcasing historical knowledge but also harm the collective memory and cultural identity of affected communities (Kamash 2018; Baird et al. 2023).

The project to replicate the arch was initiated by the Institute for Digital Archaeology (IDA), an organization dedicated to using digital tools for the preservation of cultural heritage. Their work supports broader conversations around digital restoration, especially in areas affected by political instability or conflict. The IDA recognizes the growing need for fast, repeatable methods of documenting and restoring at-risk heritage, and has worked to develop approaches that can be adapted across different regions. A major part of IDA's work consists of gathering data to map cultural sites and objects, with the goal of building a central archive of digital 3D models. This led to the launch of the *Million Image Database*, an open-source initiative aiming to collect millions of images of vulnerable artifacts and monuments.

For the site of Palmyra, which was inaccessible during the height of the conflict, the focus shifted to finding ways to protect and document it remotely. To reconstruct the arch, researchers relied on existing documentation and photographs taken by tourists, scholars, and official missions before the outbreak of conflict. These images were processed using photogrammetry, a method that extracts spatial and dimensional information from two-dimensional photographs (Remondino and El-Hakim 2006; Remondino et al. 2008; Karami et al. 2021). Specialized software was used to generate accurate three-dimensional models of the monument. This allowed the team to generate a low-resolution yet reliable digital replica of the arch. The 3D model was uploaded to open-source platforms to support documentation, education, and future restoration efforts (Fig. 5.12). This initiative highlighted how shared digital knowledge

Figure 5.12 Digital model of key structures from the archaeological site of Palmyra, including the theatre, the monumental arch, the tetrapylon, and the Temple of Bel. The spatial data were generated using photogrammetry, a method that estimates dimensional information from photographs taken on-site from multiple angles. Source: open access Palmyra 3D Model, https://sketchfab.com/tags/palmyra.

can offer a practical and symbolic response to cultural loss, even in areas that remain physically inaccessible. It also demonstrated that physical replicas, created through contemporary technologies, can carry both cultural meaning and technical innovation in tandem (Denker 2017).

To turn the digital model into a physical structure, the team opted for robotic fabrication. Digital data from the photogrammetric survey was translated into toolpaths, enabling a robotic arm to mill the components from marble. The milling process took place in Carrara, Italy, at the workshop of Torart[5], a company specializing in digital stone carving. A six-axis robotic arm was used to shape the blocks (Figs. 5.13 and 5.14), which were then assembled into a scaled replica of the arch (Fig. 5.15). The physical model was produced at one-third of the original size and served as a proof

Figure 5.13 Robotic fabrication of a block from the monumental Arch of Palmyra, 2015. The project was implemented by Torart and D-Shape, Italian companies specializing in large-scale robotic custom production. Courtesy of Torart.

Figure 5.14 Robotic milling of decorative details from Palmyra's Arch, 2015. Courtesy of Torart.

of concept for a potential full-scale reconstruction of the original 20-meter (66-foot) tall monument.

The finished arch was presented in London during the 2016 UNESCO World Heritage Week (Figs. 5.16 and 5.17), where it served as a demonstration of the potential for digital restoration to act across borders and beyond moments of crisis. The project achieved several key outcomes: it gathered visual documentation of the monument based on pre-destruction conditions, created a digital model to support future restoration, tested fabrication workflows for architectural-scale components, and produced a physical prototype to assess cost, timing, and precision. The public presentation also helped raise awareness about the value of digital archives and collaboration in the proactive preservation of cultural heritage (Kamash 2017).

In a typical digital fabrication approach to heritage conservation, especially for archaeological sites like Palmyra where artifacts are primarily made of limestone, the first intervention after damage would be anastylosis. This method focuses on reassembling the

Figure 5.15 Assembly of the arch on-site in Carrara. Courtesy of Torart.

original fragments and preserving as much of the existing material as possible, while integrating new parts to restore the overall shape. The Palmyra replica, however, followed a different path. It was not intended as a direct intervention or part of an on-site restoration effort. Instead, it was conceived as an off-site, independent

Figure 5.16 The robotically fabricated replica of the monumental Arch of Palmyra was unveiled in London on April 19, 2016, during World Heritage Week. The 1:3 scale arch stands 5.5 meters tall and weighs 11 tons. It is made from 70 cubic meters of marble and sandstone. Photo by Sara Codarin.

contribution to the preservation of cultural memory. At a time when physical access to the site was not possible and traditional restoration could not proceed, the replica functioned as a statement: that heritage can be documented, interpreted, and remembered through digital and technological means, even when conservation on location is delayed or indefinitely postponed. It was an exercise in the power of possibility, even when facing multiple obstacles beyond typical conservation effort capabilities.

The replica also draws attention to a broader issue. In situations marked by political and social instability, prompt reconstruction often becomes a priority. Yet under these conditions, routine maintenance is difficult, and long-term preservation efforts are hard to coordinate. A lack of a consistent management strategy has left many stone blocks at the site vulnerable to serious weathering, including damage from fluctuating levels of dampness, humidity, and temperature. As a response to the challenge of delayed on-site intervention, the project shows what can be achieved when conditions do not cooperate, demonstrating the potential of digital

Figure 5.17 Arch of Palmyra, detail. Photo by Sara Codarin.

tools, especially robotic fabrication, to produce components at the required level of detail independently. These parts could be used to reconstruct structural connectors and replacement elements, contributing to the safety and resilience of future restoration efforts.

Within the field of digital fabrication, subtractive processes have proven viable for this type of work. Toolpath programming allows for control over the level of detail in ornamental patterns, supporting interventions that are both reversible and visually distinct from the original. This aligns with established conservation principles, ensuring that new additions do not blur with the historic material. The programmed machines can also define the visual character of the reconstructed parts and produce simplified versions of lost decorations. In this way, the physical output becomes a formal artifact in itself, without resorting to literal replication of destroyed elements. This idea of simplification carries forward the logic of *tratteggio*, a conservation technique in fresco restoration where missing sections are reconstructed using hatched lines. Similarly, simplified three-dimensional features on the replica act as a restrained translation, signaling absence without inventing detail.

Looking further ahead, additive manufacturing methods may also offer promise. Powder-bed deposition techniques, such as the cartesian D-Shape printing technology, can produce stone-like textures that approximate the appearance of traditional materials. Envisioning a future-oriented workflow, 3D printing technologies could be deployed directly on-site, acting as mobile production units. These systems could use locally sourced sand and aggregate mixed with binding agents to fabricate custom replacement parts. This would not only reduce material transport but also help create a stronger and more authentic connection between the new inserts and the existing archaeological remains (Codarin 2016).

The arch thus served as a testbed for applying digital fabrication to heritage contexts. More than a technical exercise, it became a platform for sharing knowledge and encouraging global dialogue on how digital manufacturing intersects with heritage conservation. It also demonstrated how contemporary tools, rather than replacing traditional practices, can carry their own historical significance and contribute to the ongoing story of a monument's preservation, breathing new life into architecture that may have perished otherwise.

Case Study 3 | (re)construction

Rebuilding from Rubble with Robotic 3D Printing

The Italian startup Centauroos exemplifies how digital fabrication can be applied to architecture and construction by developing advanced systems that convert construction and demolition waste into materials suitable for 3D printing.[6] The project explores new design potentials at the intersection of material science, additive manufacturing, parametric design, and robotic construction technologies. Founded in response to the 2016 seismic events in Central Italy, where widespread damage to housing and entire urban clusters left vast territories devastated, Centauroos seeks to rethink the use of rubble, transforming it into a safe, controllable, sustainable and functional construction resource (Fig. 5.18).

The startup's design-to-manufacture process is powered by a robotic system equipped with a customized extruder for cement-based materials (Fig. 5.19). Rubble is crushed to appropriate particle sizes, reprocessed, and extruded through the robotic arm's actuator. The objective is to recycle up to 30 percent of the aggregates from debris, reducing energy consumption, material use, and production time. The extruded material mimics the characteristics of mortar sand and incorporates reclaimed aggregates. These can be used to 3D print architectural components in various formats, which are then assembled to form finished

Figure 5.18 Centauroos' core concept: transforming post-earthquake and post-demolition construction rubble into a 3D-printable material for robotic fabrication. Courtesy of Centauroos.

Figure 5.19 Custom-built actuator for robotic 3D printing using reclaimed aggregates for off-site material production. Courtesy of Centauroos.

Figure 5.20 *P.System*, 2024. Urban furniture designed for reconfiguration, 3D printed using recycled demolition waste. Courtesy of Centauroos.

elements ranging from wall modules, façade cladding, or urban furniture (Fig. 5.20).

Centauroos engages with urban restoration, particularly in rural villages and historic clusters, introducing performance-driven built environments in contexts of post-earthquake reconstruction or demolition. These settings provide ground for architectural experimentation, activating a regenerative approach that reintegrates demolition waste into contemporary construction workflows. The same methodology supports the 3D printing of anti-seismic structures, using rubble as a byproduct reinserted into a material stream managed through a computationally guided, robotically executed design process.

Among Centauroos projects, *Evolving Stonework* (Fig. 5.21) investigates historic masonry textures in rural buildings to study the interface between traditional constructions and new materials. The cementitious extrusion mix can be tailored per case to meet different strength and performance criteria, enabling prefabrication of

Figure 5.21 Façade studies exploring textures and the interface between historic construction techniques and a new 3D-printed tectonic system. Project presented at the 2025 Venice Architecture Biennale, curated by Carlo Ratti. Courtesy of Centauroos.

blocks that are either monolithic or segmented, depending on design needs. The traditional aesthetics of brickwork—often restored using *cuci-scuci* (stitch-and-cut) techniques—are reinterpreted through the horizontal layering of extruded material, creating a novel architectural stratification. This project reflects a broader vision pursued by Centauroos: a paradigm of collective, collaborative construction where humans and machines operate within the same workspace (Ratti 2025). Through digital additive fabrication, localized production, and material reuse, the company advances a form of digital craftsmanship rooted in both tradition and innovation. The design process is also tested in augmented reality to simulate potential on-site operations, including damage assessment, structural reinforcement, and additive construction (Fig. 5.22).

This approach also resonates with emergency contexts; during the 2016 earthquake in Central Italy, for instance, robotic systems were used in post-disaster search and rescue within unstable buildings. Such scenarios emphasize the urgency of integrating innovative technologies into construction workflows, particularly in environments where structural safety is uncertain and human access may be hazardous.

Figure 5.22 Human–robot collaborative workspace controlled through augmented reality. Courtesy of Centauroos.

The research further invites reflection on how this approach could scale to address broader challenges in construction waste management. Additive manufacturing presents promising strategies for reducing material consumption and accelerating building timelines. In Europe alone, over 510 million tons of construction and demolition waste are generated annually, with processing costs reaching 45 billion euros. Looking ahead, one scenario envisions geolocated repositories of construction waste, regularly updated to track material availability, that would supply printable feedstocks to decentralized manufacturing hubs. This distributed model could support on-site robotic fabrication, transforming waste directly into architectural components where they are needed most.

Case Study 4 | repair

On Site Robotics: Traditional House of the Future

In 2023, Lidia Ratoi and John Lin, researchers from the University of Hong Kong, developed the *Traditional House of the Future*,[7] a residential project built in Guiyang, China (Figs. 5.23 and 5.24). This work brought together their previous research and explored a space between robotic fabrication and the social restoration of rural areas. The project is part of a government initiative aimed at renovating hundreds of old wooden houses that have deteriorated over time due to the consequences of shifting social dynamics in southeast China. The process unfolded in a village that was on the verge of abandonment, where residents had begun relocating to nearby settlements consisting of generic concrete buildings lacking modern amenities. The project's aim was to reverse this trend by prototyping a full-scale construction that employed advanced technologies, offering both a financial incentive and a new center of attraction for

Figure 5.23 *The Traditional House of the Future*, 2023, in Guiyang, China, a full-scale proof of concept integrating robotic 3D printing with vernacular timber construction to reactivate rural housing through technological and cultural continuity. Courtesy of Lidia Ratoi.

Figure 5.24 *The Traditional House of the Future.* Project overview. Courtesy of Lidia Ratoi.

the community. The house was envisioned as a replicable model of best practices that could transcend construction itself and serve as a catalyst for restoring a positive social environment. This ambition required a careful understanding of local construction traditions and the cultural landscape. The research started with an investigation into vernacular building practices, focusing on how self-builders adapt their homes in response to the challenges of rapid urban growth. These adaptations produce evolving architectural forms, where traditional methods blend with newer techniques, resulting in unique typologies marked by a distinct sense of craftsmanship. The *Traditional House of the Future* drew directly from these local strategies, integrating them with robotic construction methods to establish a connection between rural China and emerging technological approaches, particularly through an opportunity to combine in-depth knowledge of traditional wooden detailing and large-scale 3D printed artifacts.

One existing two-story house in the village was chosen as a test bed. The process was initiated by scanning its geometry and wooden structure to generate a digital model, which served as a reference for exploring how 3D-printed components could be integrated with the existing building. After the data capture phase, a design-to-fabrication strategy was defined to reorganize the ground floor layout. This involved adding new volumes that were 3D printed in situ (Fig. 5.25), allowing the existing structure to expand and adapt to contemporary spatial needs. For this reason, and to prepare the space for the machinery to operate on site, the original structure was carefully dismantled. Its wooden elements were cleaned, sorted, and reconditioned for reassembly once the printing phase had been completed. This method was possible because traditional rural Chinese wooden houses are constructed entirely through a joinery method. Without the use of nails or hammers,

Figure 5.25 Industrial robotic arm transported on site for concrete 3D printing. Courtesy of Lidia Ratoi.

these buildings can be taken apart and put back together within a single day.

The project progressed through the use of industrial robotic arms equipped for concrete printing and extrusion, which were deployed directly on site (Fig. 5.26). The construction followed a process of ongoing adaptability, with design and fabrication phases overlapping and evolving in parallel until the building was fully realized. This was especially evident when the wooden structure had to be rebuilt around the printed concrete components

Figure 5.26 On-site 3D printing with concrete robots reconfigures the ground floor of the dismantled timber structure, aligning contemporary spatial needs with vernacular form. Courtesy of Lidia Ratoi.

Figure 5.27 Reassembly of the cleaned and catalogued timber components around the newly robotically printed curvilinear concrete elements. Courtesy of Lidia Ratoi.

(Fig. 5.27). Craftspeople responded with inventive solutions to shape the timber framework around the geometry of the 3D printed walls to reconstitute the house (Fig. 5.28). These new walls are curvilinear elements designed not only as boundary conditions but treated as dynamic spaces that become part of

Figure 5.28 Craftspeople shape the timber envelope in dialogue with the printed walls, merging high-tech fabrication and material intelligence with human intuition and adaptability. Courtesy of Lidia Ratoi.

Figure 5.29 The final structure weaves together printed elements that serve as walls, planters, and furnishings, defining spatial thresholds while enabling a new model for social and architectural continuity in rural settings. Courtesy of Lidia Ratoi.

daily life (Fig. 5.29). The 3D printed parts were developed in an organic and continuous way, so in different areas of the house, they take on multiple functions: they become planters, stairs, and wrap around service spaces such as kitchens, bathrooms, and entryway thresholds. This weaving of inside and outside was a deliberate formal strategy that reinforced the building's role as both a place of shelter and a space for community life, actively engaging its users with shared experiences.

The *Traditional House of the Future* represents a prototype that captures the shifting ways of life in rural China (Fig. 5.30). Positioned between categories, not fully traditional or modern, rural or urban, the project reflects how residents navigate thresholds between indoor and outdoor living. The design builds on these habits, offering an updated model of the wooden house that accommodates present-day needs and adaptable spatial arrangements. Local residents were actively involved throughout the construction

Figure 5.30 Local community at Guiyang, China. Courtesy of Lidia Ratoi.

process, shaping a participatory model that merged robotic on-site fabrication with wood craftsmanship. Rather than treating technology as a replacement for traditional methods, the project presents it as a social amplifier, reinforcing cultural practices and addressing sustainability in social and cultural terms as much as environmental or technical ones. In this context, restoration addressed not destruction but abandonment: the gradual disappearance of the elements that once sustained village life. The addition of on-site fabrication materials supports and strengthens the original wood structure, sustaining longevity. The project reimagines ancient dwellings not as fixed artifacts, but as flexible, evolving structures capable of responding to the present while preserving living heritage, building continuity through change.

Case Study 5 | repair

Robotics for the Architectural Reuse of Salvaged Building Materials

A research trajectory that expands the role of digital tools relating to existing conditions, and having a focus on conservation, repair, reuse, and adaptive transformation, has been developed by Malgorzata Zboinska at Chalmers University of Technology. In particular, the project *Digital Tool Integrations for Architectural Reuse of Salvaged Building Materials*, conducted with Frederik Göbel, presents a framework for facilitating the reuse of salvaged materials, especially those irregular in nature such as timber, through advanced computational design and robotic workflows using biobased materials[8] (Zboinska and Göbel 2025). In this project, additive manufacturing is applied to join broken elements, allowing damaged pieces to be reassembled and put back into use.

The premise of the work stems from an urgent concern regarding the massive amounts of waste and emissions generated by the construction industry and its demolition practices. Architectural reuse faces significant challenges due to the diversity and damage typical of reclaimed materials, often disqualifying them from reuse. However, the methodology proposed in this research could readily be extended to the restoration of damaged building components or artifacts by introducing a layer of design agency enhanced with advanced digital tools. The project addresses these challenges by proposing a digitally integrated process that enables damaged materials to be efficiently repaired and reintroduced into new construction with minimal waste. Through a combination of digital capture techniques, such as photogrammetry, 3D scanning, and robotic vision, the work supports the development of a fully circular design process. In this specific study, the reclaimed architectural elements were timber floorboard fragments recovered from a demolition site,

Figure 5.31 Reclaimed wooden floorboards used in robotically-enabled experimental reuse interventions, shown as textured mesh models captured through photogrammetry. Courtesy of Malgorzata Zboinska.

each exhibiting different types of damage (Fig. 5.31). The 3D printing interventions used to repair and adapt these elements employed filaments composed of wood and hemp biomass.

The method is structured in three stages: digital data capture, data processing, and data harnessing through the data-driven generation of robotic fabrication sequences. The first stage uses three capture methods: 2D photography, 3D photogrammetry (Fig. 5.32), and structured-light 3D scanning with a robot-mounted camera

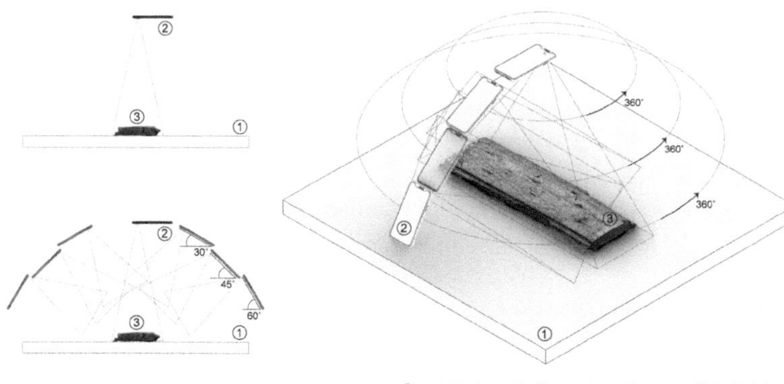

Figure 5.32 Setups for 2D image capture and photogrammetry. Top left: single image capture using a smartphone camera. Bottom left and right: photogrammetry setup with overlapping image series taken at multiple angles, distances, and panoramic ranges. Courtesy of Malgorzata Zboinska.

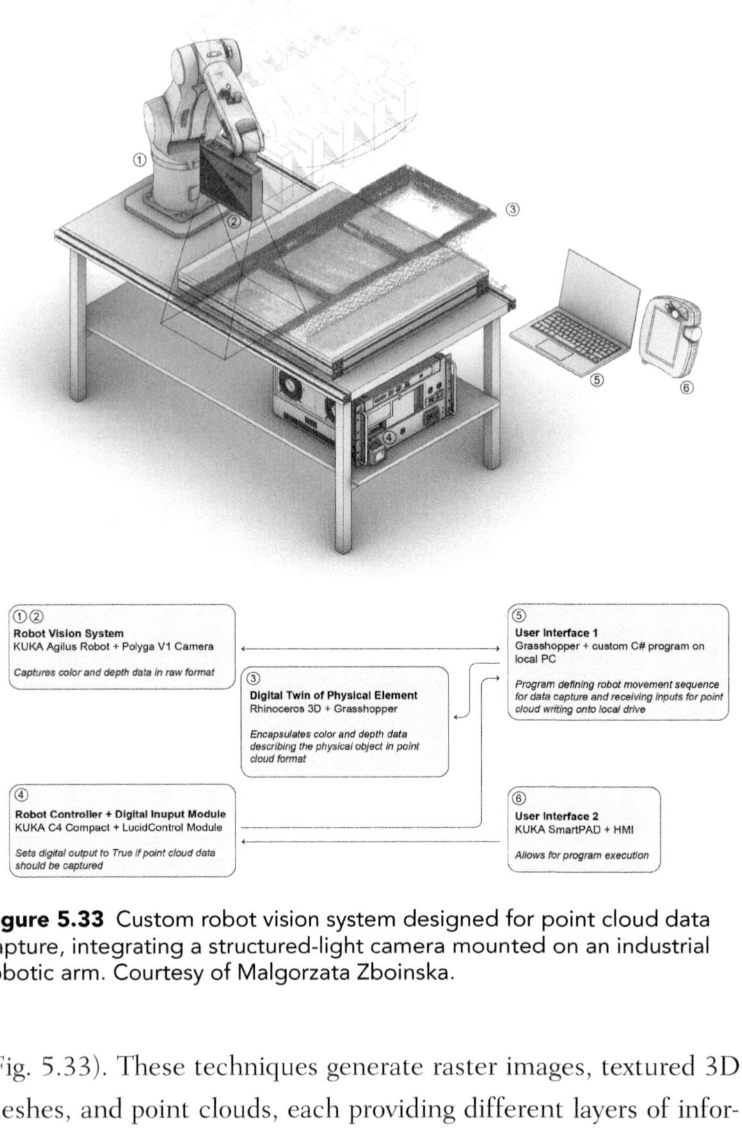

Figure 5.33 Custom robot vision system designed for point cloud data capture, integrating a structured-light camera mounted on an industrial robotic arm. Courtesy of Malgorzata Zboinska.

(Fig. 5.33). These techniques generate raster images, textured 3D meshes, and point clouds, each providing different layers of information on component geometry, texture, and spatial characteristics. This phase supports the analysis of individual wooden fragments, allowing the identification of defects and features to inform case-specific robotic interventions. In the second stage, the captured data is filtered, clustered, and used for feature extraction. Machine learning algorithms and computer vision techniques read the 2D images to detect knots, cracks, and grain patterns. Point cloud data is processed to isolate features based on color, brightness, and spatial data. Mesh data provides topological surface information for volumetric operations. The final stage translates this processed data into robotic toolpaths. Initial experiments used 2D image data to create decorative surface textures, with print paths generated through growth pattern algorithms informed by wood grain orientation. Other workflows identified indentations in damaged wood, generating textured repairs with Voronoi-based infill, or layered surface coatings that responded to discoloration. The final set of experiments focused on volumetric additions, restoring broken boards with custom-fitted extensions (Fig. 5.34) and designing

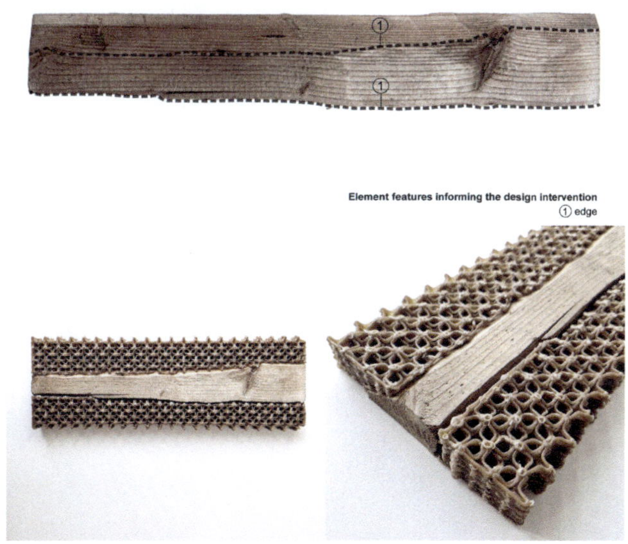

Figure 5.34 Design outcome from one of the iterations. A reclaimed wooden floorboard with an irregular form [I think these indications are based on a previous image organization] is restored to a standard shape by adding two robotically 3D printed extensions. Print paths were generated through volume voxelization, using geometric surface data extracted from the broken edges via Boolean operations. Courtesy of Malgorzata Zboinska.

interlocking joinery pieces to assemble uniquely shaped reclaimed elements (Fig. 5.35).

As the authors note, these examples demonstrate the potential for component recombination into new configurations using digitally designed hybrid material systems. Additive fabrication was selected over subtractive methods such as milling, as it avoids generating new waste and can accommodate irregular geometries with greater

Figure 5.35 Design outcome from one of the iterations. Two reclaimed wooden floorboards are assembled using a custom biobased, robotically 3D printed interconnection piece, designed with a tween curve method informed by mesh sectioning data derived from photogrammetry. Courtesy of Malgorzata Zboinska.

flexibility. The research highlighted how each type of digital data offers specific advantages. Raster images are quick to capture and well-suited for surface pattern recognition, though they lack depth information. Point clouds provide depth and spatial context but are limited in resolution and lack connectivity between points. Meshes capture full surface topology, enabling volumetric operations, but make it difficult to isolate fine surface features. Combining these data sources enables holistic digital representations of reclaimed components, supporting highly customized design strategies and robotic fabrication sequences. Regarding real-world applications, the research outlines both on-site and off-site deployment scenarios, aligned with the developments of the Fourth Industrial Revolution and the ongoing digitization of construction processes. On-site approaches include the use of mobile scanning units, drones, or compact robots capable of scanning and printing in difficult or constrained environments. Off-site strategies focus on robotic repair facilities, equipped with scanning stations, 3D printing cells, and labeling systems to support prefabrication workflows.

From an environmental perspective, the proposed framework extends the lifespan of salvaged materials to reduce construction-related emissions. Robotic 3D printing, particularly when used for non-standard components (Zboinska et al. 2022), has been shown to outperform traditional construction methods in both carbon footprint and material efficiency. Reuse strategies become impactful when paired with bio-based materials, which further reduce greenhouse gas emissions. Timber, for example, is a sustainable material due to its carbon sequestration capacity, and robotic repair techniques can significantly expand the quantity of reclaimed wood that can be put back into use. The digital framework presented here supports adaptive, data-driven fabrication that empowers new architectural assemblies by respecting the unique characteristics of each component.

While rooted in material reuse, this work also opens a line of inquiry about digital tools informing restoration practices, particularly those requiring reintegration of material losses, or *lacunae*, in architectural elements. Such interventions can be understood at a broader scale, where responding to absence becomes both a critical and creative act. A nuanced reading of existing constructions is required, especially in cases where time has irreversibly altered the form, as seen in architectural ruins (Dalla Negra 2017). These material losses reshape the original composition, and any new addition must respond thoughtfully to that transformation. In line with established restoration principles, the integration of new material should remain legible and clearly distinguishable from the original fabric, while still maintaining visual and spatial coherence (Brandi 1963). The ability to interpret existing decorative patterns or compositional logics can guide meaningful reintegration, especially when such interventions need to restore a component's functional role. Workflows that support these adaptable and scalable approaches, such as the technical repair of damaged elements and the opportunity for creative recombination, contribute to an evolving model of architectural repair that is both technologically informed and environmentally responsive.

Case Study 6 | repair

Robotic Strategies for Repairing 3D Printed Biocomposites

Advancements in material science and the emergence of novel fabrication processes have not only introduced new possibilities for architectural expression but also raised critical questions regarding the life cycle of these materials and relevant strategies to extend their functional lifespan. In this context, the Centre for Information

Technology and Architecture (CITA) in Copenhagen has positioned itself as a leading research environment specializing in biobased materials, robotic fabrication, and circular design methodologies. Their recent work proposes a robotic approach to the repair of weathered biocomposite panels, and demonstrates how robotic 3D printing can extend material longevity through a design-for-repair methodology[9] (Chiujdea et al. 2024). This research reframes repair as a generative and anticipatory act, embedding it within the initial design process rather than treating it as a post-failure response. By connecting diagnosis, material behavior, and toolpath generation into a digitally coordinated, closed-loop system, CITA grants biocomposite architecture as a new temporal dimension that acknowledges, and even incorporates, decay as part of an ongoing material dialogue.

Biobased materials such as cellulose and lignin-based polymers have gained particular traction because of their renewable origins and potential for carbon sequestration (Ramsgaard Thomsen et al. 2023; Rossi et al. 2023; Thomsen and Tamke 2022). However, their hygroscopic nature, the way they swell, shrink, and degrade in response to moisture and temperature, creates performance challenges, especially in outdoor environments. Traditional repair frameworks such as design-for-disassembly or even restoration practices are designed for standard materials and modular systems, but they do not translate well to biocomposites. The continuous and anisotropic, meaning directionally dependent, character of these materials calls for alternative strategies that respond to their inherent tendency to change over time, treating transformation as an anticipated design feature rather than a defect. To address the challenges of material degradation, the research group proposes a design-for-repair workflow that spans from robotic manufacturing to robotic repair.

The building component at the center of this investigation is a panel conceived as a multi-layered façade element, composed of

Figure 5.36 Robotically fabricated biocomposite panel produced through 3D printing. Courtesy of CITA.

three functional strata: a structural base layer fitted with joinery, a sacrificial overhang layer designed for weather protection, and a connective layer that bridges the two (Fig. 5.36). A generative algorithm defines the toolpaths for each layer, taking into account extrusion limits, nozzle diameter, and fiber content. Each layer is color-coded through the selection of specific cellulosic fibers, such as wood flour, bark, and cotton, adding a functional layer of legibility that makes the effects of weathering visually traceable. Once fabricated, the panels are installed outdoors and subjected to real-world climatic conditions from July to October in Copenhagen. As anticipated, signs of degradation appear across multiple parameters: panel edges recede due to shrinkage, overhangs lose their geometric definition, and some panels develop cracks (Figs. 5.37 and 5.38). Discoloration and fiber detachment further illustrate the vulnerabilities of biocomposite structures under environmental stress. In response to these changes, the researchers introduce a multi-step diagnostic process using photogrammetry. Each panel is scanned before and after exposure,

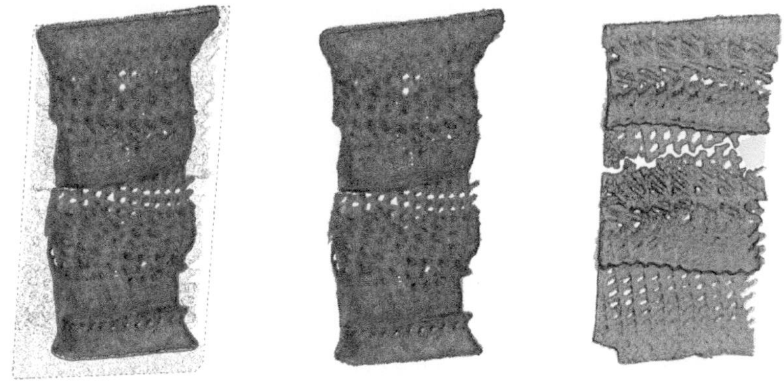

Figure 5.37 Digitally scanned data used for diagnosis. Left: boundary difference indicating receding edges. Center: clustering of the overhang layer showing loss of definition. Right: boundary proximity analysis detecting cracks and missing parts. Courtesy of CITA.

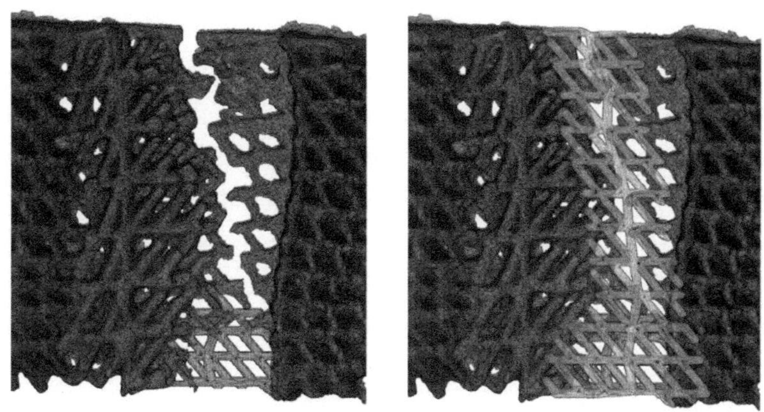

Figure 5.38 Diagrams based on digital scan data, illustrating the designed patch geometry used to fill missing areas (left) and the stitching geometry applied to reconnect the two sections of a cracked panel (right). Courtesy of CITA.

allowing for the production of high-resolution mesh reconstructions aligned through fixed registration markers. The resulting digital models offer a dataset that support comparisons between original and weathered conditions.

The diagnostic phase identifies three main types of deterioration: receding edges, loss of definition, and cracks. Each condition

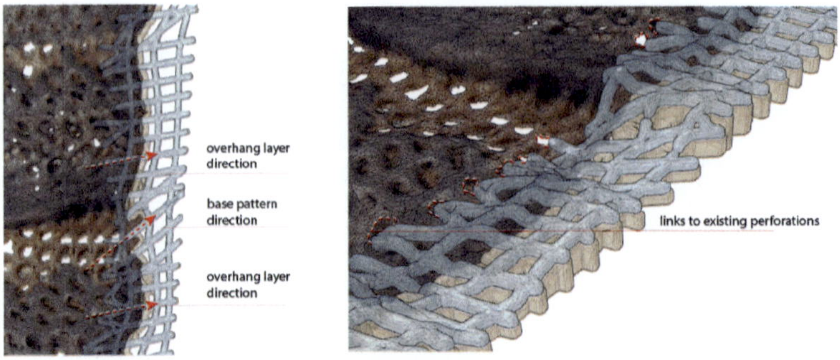

Figure 5.39 Diagrams illustrating the directionality of the pattern (left), the layer differentiation (center), and the connection to existing perforations in the weathered state mesh (right). Courtesy of CITA.

is addressed through a dedicated repair strategy. In particular, when cracks cause the panel to split, a patch and stitch method is employed. A patch is used to fill the missing material between the separated edges, and a stitching layer is printed across the fracture to reconnect the broken parts, restoring both continuity and structural integrity. These repairs are printed conformally, meaning directly following the surface geometry, the weathered panels using a UR16e robotic arm and a custom extruder setup (Figs. 5.39 and 5.40). Unlike flatbed printing, conformal 3D printing requires the toolpath to follow the curvature of an irregular surface, adapting dynamically to the geometry of each panel. The repair paths are

Figure 5.40 Conformal robotic 3D printing process used to repair cracked biocomposite panels. Courtesy of CITA.

designed as material-specific interventions, informed directly by the data captured during the diagnostic phase.

Repairing rather than replacing panels reduces waste, and the use of biodegradable, renewable inputs helps lower the carbon footprint of construction. More significantly, the intentional integration of diagnostic and repair parameters into the initial design enables multiple cycles of deterioration and renewal. In this model, the design becomes a living process, capable of recognizing and responding to its own breakdown. While we are still in the early stages of developing this approach, there is a growing sensibility in material research that embraces decay and embeds intelligence into fabrication workflows. This emerging attitude allows repair actions to be anticipated, systematized, and potentially replicated across a range of found conditions. Historically, long-established materials such as brick have shaped clear repair practices, like the *cuci-scuci*, stitch-unstitch, a disassembly-and-reassembly methods used to restore already damaged masonry. In contrast, biopolymer prints represent a new class of architectural elements that are not defined by human-scale dimensions, can be either discrete or continuous, and are fabricated at varying scales with diverse assembly patterns. In the case of this experiment, the repair process seems less like *cuci-scuci* and more like a simple *cuci*—the printing of an additional material layer that overlays the existing form, resembling the act of sewing a seam with a machine. From these new materials and fabrication logics, new methods of repair and terminologies are emerging, laying the foundation for a revised methodological framework. Here, repair becomes not only restorative but also formative (Thomsen et al. 2024). Reframing repair as a creative and ecological act (Oropallo 2019) provides a meaningful direction for decay-aware construction supported by digital craft, especially in an era marked by an environmental urgency to preserve both buildings and resources.

Case Study 7 | aesthetic language

Digital Gothic: Robotic Carving for Bespoke Formwork

Historic architectural languages serve multiple functions: they carry the cultural and construction legacies of the past while acting as a formal and conceptual archive that can be sampled in the development of new architectural expressions. An example that uses historical references and aesthetic values as generative inputs, processed through contemporary tools, is *Variable Carving Volume Casting*, a research project developed at MIT.[10] The project proposes a robotic method to produce customized formwork to cast unique volumetric architectural elements that are inspired by the formal logic of Gothic craft traditions. Carving, a technique historically a manual craft, is reinterpreted here as a digitally driven fabrication strategy. A multi-axis robotic arm equipped with a hot-knife is used to rapidly carve expanded polystyrene (EPS) foam into negative molds. These are then used to cast geometrically complex forms made of glass fiber reinforced gypsum, and result in a repeatable, low-waste, and materially responsive workflow for bespoke architectural production (Clifford et al. 2014).

The work responds to a critique of prevailing assumptions in current discourse regarding digital fabrication and casting. While architecture has long explored volumetric materiality, modern construction tends to rely on sheet materials and standardized components. Current methods for casting complex geometries, such as CNC milling of molds or flexible formwork with pneumatic actuation, either produce substantial material waste or are limited in their geometric resolution due to tool size constraints. The research team identifies this as a tension between digital variation and analog constraint, where true geometric freedom is often sacrificed in favor of repeatability or efficiency. The project aligns with what is described as a *digital gothic* sensibility, drawing conceptually from John

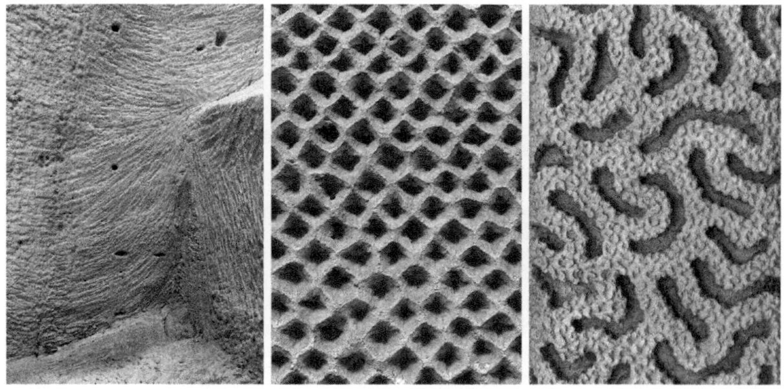

Figure 5.41 Carved stone interiors and facades displaying distinct tool mark patterns. In some cases, markings follow the geometry of excavation, revealing directional tool logic. In others, chiseled textures form deliberate surface treatments, such as inverted pyramids or vermiculation, illustrating efficient techniques for adding visual and tactile depth to architectural elements. Courtesy of Brandon Clifford.

Ruskin's notion that gothic architecture was determined by the tools and methods of the maker, whereas classical traditions governed form by abstract ideals (Ruskin 1900). In this approach, robotic tooling is not limited to only achieving precision but also becomes an expressive and reciprocal act that restores agency to the process of making. Tool marks are no longer treated as imperfections but are instead embraced as evidence of fabrication logic, much like the visible chisel patterns intentionally left behind by medieval builders (Fig. 5.41). The use of the robotic hot-knife enables both speed and geometric specificity, while also celebrating the physical trace of the tool in the final cast surface.

The study uses the typological element of the column as a basis for investigation (Fig. 5.42), because it provides a means to examine bundling, fluting, and other historical variations of material aggregation. Ancient and classical column traditions, from Egyptian reed bundles to the twisting Solomonic spirals of the Baroque, are revisited to align with a long lineage of tool-informed design. Instead of

Figure 5.42 Historical examples of fluting and bundling in column design. Initially derived from bundled materials that cinched and bulged when tied, these forms were later interpreted in stone, demonstrating the enduring influence of material behavior on architectural expression and ornamentation. Courtesy of Brandon Clifford.

replicating historical motifs, the research draws from their underlying logic and responds to contemporary fabrication constraints. The column's fluted surface, for example, becomes a site for negotiating hot-knife geometry, sweep radius, and surface resolution. Two versions of the hot-knife tool were prototyped: an off-the-shelf version adapted for robotic control and a custom-mounted variant that provided a larger cutting radius (Fig. 5.43). These tools introduced

Figure 5.43 Robotic end effector prototypes for foam carving: an off-the-shelf hot-knife with a J-shaped blade (left), and a custom-mounted variant with a larger curve (right), each supporting different carving strategies. Courtesy of Brandon Clifford.

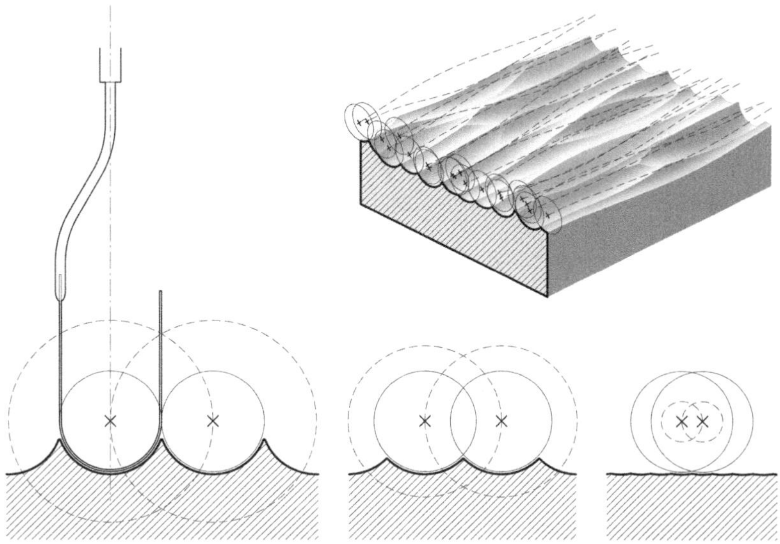

Figure 5.44 Digital diagram of robot tool pathing. Courtesy of Brandon Clifford.

geometric constraints and made obvious aesthetic affordances. Unlike CNC milling, where tools can change direction freely, the hot-knife blade has a fixed direction and turning radius that determines how the robot can move. Toolpaths must be carefully calculated to avoid over-melting and ensure consistent cutting depth, balancing the minimum and maximum step-over distances. The result is not a pixel-perfect smooth surface but a controlled roughness that mixes the layered striations of traditional craft and contemporary robotic logic (Figs. 5.44 and 5.45).

Digital models were generated using parametric tools in Rhino and Grasshopper, allowing curves to modulate across the column's surface. The resulting molds were carved from EPS foam blocks and sealed with latex membranes before casting (Fig. 5.46). Initial mold release strategies, such as waxes and oils, proved inconsistent, so the team developed a vacuum-formed latex sheet system that produced smooth, detailed surfaces while protecting the mold

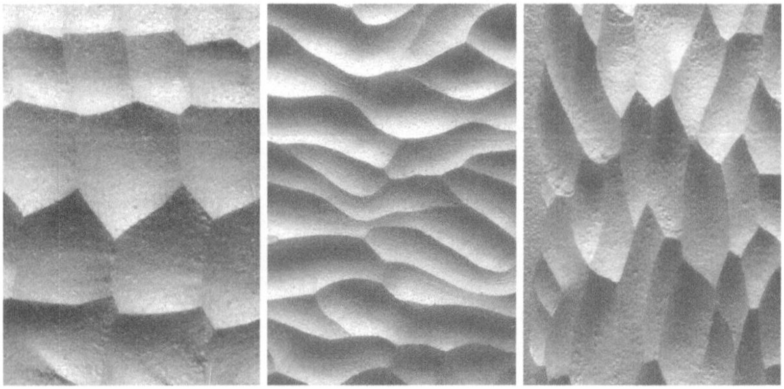

Figure 5.45 Multiple toolpath prototypes developed to test carving strategies and surface effects across varying sweep angles and material responses. Courtesy of Brandon Clifford.

Figure 5.46 Robotic arm with hot-knife assembly carving expanded polystyrene foam to create a negative mold, shown alongside a finished cast column produced from the same formwork. Courtesy of Brandon Clifford.

for reuse. Variations in latex thickness were tested, revealing trade-offs between edge sharpness and polish, and highlighting how a casting resolution can be tuned through material choice and membrane behavior. The cast elements in the study were produced using gypsum reinforced with chopped strand fiberglass (Fig. 5.47) and

Figure 5.47 Robotically carved foam formwork and the resulting cast element, each reflecting the specific toolpaths and surface articulation produced through robotic carving. Courtesy of Brandon Clifford.

to assemble the columns, two halves were cast separately, then aligned and permanently bonded using expanding urethane foam. The assembly process was as much about negotiation between surfaces as it was about structural joining. Imperfect interfaces were accepted and worked with rather than machined away, further reinforcing the project's commitment to craft-informed fabrication.

Rapid carving of large-scale molds using a robotic hot-knife supports unique production, serial in nature with minimal waste, and avoids the high costs and time demands of traditional subtractive machining. The process is scalable, materially efficient, and receptive to formal variation. The proposed method also introduces a flexible workflow capable of responding to specific site conditions, design intent, and material behavior where the hot-knife can function not only as a shaping tool but as a medium for rethinking casting-based design strategies. This can be applied in contemporary re-construction scenarios, where volumetric additions contribute to the restoration of historic buildings. In these cases, historic formal patterns can inform robotic toolpaths and surface articulation. When aligned with both volumetric thinking and traditional knowledge, robotic fabrication has the capability to generate new languages of making that are grounded in responsiveness, variability, and embodied feedback. The result is a reconciliation of historical craft through contemporary automation, a model for robotically customized casting that is materially sensitive, geometrically adaptable, and contextually aware.

Case Study 8 | aesthetic language

Biobased Coating Using Robotic 3D Printing

Current challenges in the construction industry, particularly those concerning material reuse, renovation, and the preservation of material culture, have prompted innovative experimentation in the realm of environmental sustainability and the development of novel material systems. In this context, Malgorzata Zboinska and her team explore the use of robotic 3D printing to apply biobased coatings onto architectural timber elements, using microfibrillated cellulose hydrogel[11] (Zboinska 2024). This work opens up new possibilities for restoration, where patterned material deposition enables volumetric addition or surface reintegration, reworking damaged and aged elements (Rudin et al. 2022). These operations suggest a contemporary language of repair, shaped by material behavior, design intention, and digital tooling. Differently from the examples of repair, this project operated on the surface of existing objects, reinforcing and renewing the outer layers without altering the overall mass of the element.

The biobased coating system used for robotic 3D printing is derived from renewable biomass byproducts such as forestry and agricultural waste, aligning with broader efforts to establish circular practices in architectural production. While much research has focused on biobased materials as independent building components, this project investigates their integration with existing substrates, in this case timber, at an architectural scale and with attention to visual and cultural continuity. The coatings are thus positioned as both a sustainable material strategy and a tool for preservation, aesthetic consideration, and adaptive repair.

The project follows a research-by-design methodology that uses prototyping to build knowledge through direct experimentation (Zboinska 2021). Architectural details from buildings in Gothenburg, Sweden, such as wall cladding, railings, friezes, and sculpted beams, served as the basis for the investigation (Fig. 5.48). These elements were replicated as eight mockups, fabricated from spruce and pine, to explore how the coatings interact with different species of wood (Fig. 5.49). Each was selected to reflect common

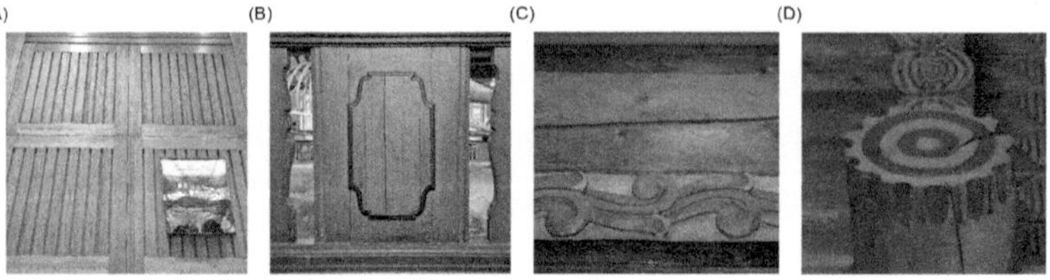

Figure 5.48 Architectural details sampled from existing buildings in Gothenburg, Sweden. (A) Wall cladding at Bergsjöns kyrka (1975, architect: Bo Cederlöf). (B) Mezzanine railing at the Natural History Museum (1923, architect: Ernst Torulf). (C) Frieze at Masthuggskyrkan (1914, architect: Sigfrid Ericson). (D) Sculpted beam at Masthuggskyrkan. Courtesy of Malgorzata Zboinska.

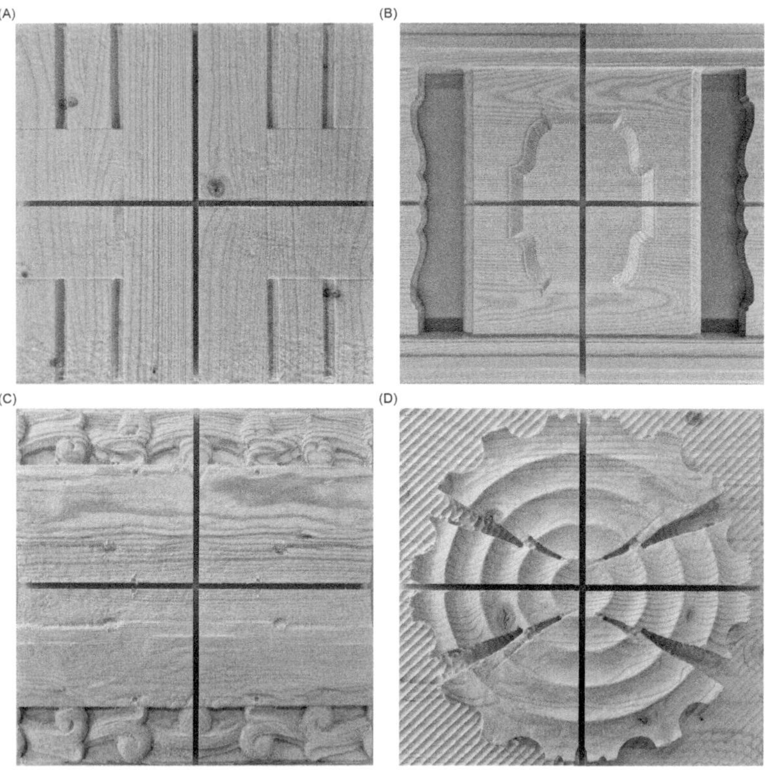

Figure 5.49 CNC-milled mockups of architectural detail fragments based on elements sampled from existing buildings. (A) Wall cladding fragment. (B) Mezzanine railing fragment. (C) Frieze fragment. (D) Sculpted beam fragment. Courtesy of Malgorzata Zboinska.

conditions of deterioration and to test the potential for protective surface enhancement. The pattern designs, developed through parametric modeling, drew from the architectural language of each source, referencing symbols such as ichthys motifs[12] and dragonfly wings. The coatings themselves consisted of pigmented cellulose hydrogel blends, applied with a robotic arm and equipped with a custom extrusion system (Fig. 5.50). To assess how surface preparation influences performance, three types of precoating layers were tested. After printing, the models were left to dry in ambient

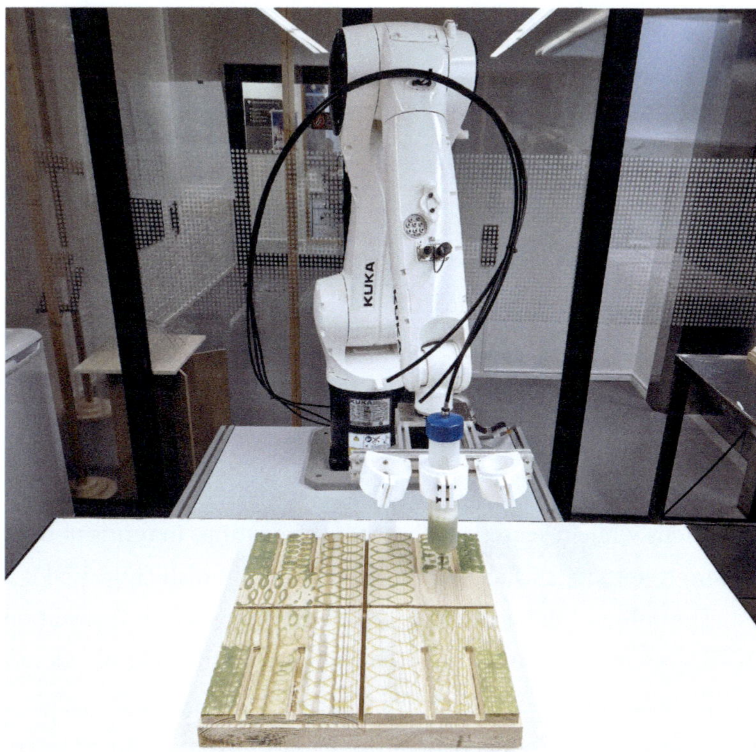

Figure 5.50 Digital fabrication setup for the prototyping experiments, showing the robot cell and the custom-built end-effector designed for hydrogel extrusion. Courtesy of Malgorzata Zboinska.

Figure 5.51 Architectural features of biobased coatings at the macroscale, with final appearances closely reflecting the initial digital pattern designs. Patterns include the ichthys motif on the wall cladding fragment, the Voronoi structure on the mezzanine railing fragment, the flowline pattern on the frieze fragment, and the circular arrangement on the sculpted beam fragment. Courtesy of Malgorzata Zboinska.

conditions while researchers observed the coatings in terms of their pattern retention, surface bonding, and overall finish (Fig. 5.51).

The robotic toolpathing embraces an approach to restoration that leaves intervention visible. During drying, the coatings developed subtle surface irregularities, including curling and bridging. These features introduced material variation that contributes to the expressiveness of the restored surfaces, not as flaws but as intentional characteristics. The results resonate with craft-based traditions, where material behavior is considered part of the outcome. Restoration in this context becomes a conversation between historical matter and contemporary intervention, moving beyond the idea of simply returning to an original state. Color changes during drying additionally proved important when shaping how interventions were perceived. Some substrates absorbed pigments unevenly, causing hue shifts or softened edges, while others allowed the printed patterns to remain sharply defined (Fig. 5.52). These

Figure 5.52 Unexpected architectural features of biobased coatings at the mesoscale, emerging during the drying process and specifically exemplified in selected mockups. (A) Edge rounding. (B) Path curling. (C) Edge irregularity. (D) Path merging. (E) Color blending. (F) Discontinuities and ruptures. (G) Detachment. (H) Bridging. (I) Edge curling. Courtesy of Malgorzata Zboinska.

behaviors added another layer of complexity to the visual language of the restored surfaces and reinforced the role of materiality in shaping design outcomes (Zboinska and Dumitrescu 2021).

The study situates its findings within broader heritage conservation concerns. As stated by the author, timber is a widely used and culturally significant material present in structures ranging from Scandinavian stave churches to Asian temples and Polynesian marae. In many of these contexts, conventional synthetic coatings have been used to protect surfaces, yet such products often block moisture exchange and result in accelerated deterioration. In contrast, cellulose-based coatings remain breathable and compatible with timber, offering a promising alternative for conservation and repair that works within the material's natural behaviors and creates possibilities for both historic and contemporary practices. Through aesthetic customization, structural repair, and surface protection within a single workflow, the research presents restoration as a creative mode of engagement and an active part of the building's ongoing life cycle. This type of work contributes to rethinking how buildings are maintained and repaired while establishing a basis for future exploration across material science, digital craft, and conservation as an active and thriving design territory.

Case Study 9 | aesthetic language

Design via Additive and Subtractive Robot Behavior

Artisanal practices, grounded in a maker's material intuition, embodied skill, and iterative gesture, serve as a repository of knowledge for structuring machine-driven fabrication processes and their efforts to move beyond automation toward more expressive and responsive forms of making. In support of these goals, the *Viscera/L*[13] project, led by Robert Stuart-Smith at the Autonomous Manufacturing Lab at the University of Pennsylvania,[14] presents a strong investigation into robotic fabrication that merges additive and subtractive processes within a single sculptural workflow. The project does not prioritize the replication of identical parts or precision as an isolated objective. Instead, the focus becomes how robotic tools can engage matter through tactile, adaptive interactions. *Viscera/L* uses clay as a primary medium and reconsiders how design can draw on historical and aesthetic precedents to develop a vocabulary of form that is both ornamental and structural, while being authored uniquely through iterative material interaction (Stuart-Smith et al. 2022).

The project utilizes the long history of architectural ornament and material manipulation, particularly that of the Roman Baroque period where sculptural qualities were inseparable from architectural space. Drawing on this lineage, the project references art historian Heinrich Wölfflin's descriptions of Bernini's work, where it can be seen that form emerges from painterly gestures and is further animated by light and shadow. In Wölfflin's analysis, parts dissolve into a more atmospheric whole, giving architecture a sense of vitality that is not dependent on symmetry or proportion, but instead on movement, texture, and depth (Wölfflin et al. 1967). These ideas find a modern contemporary translation in *Viscera/L*, where the robot becomes a participant in the shaping of form rather than an executor of predefined instructions.

Fabrication moves away from previous approaches where predefined, unchanging toolpaths—and fixed movement instructions are given to a robot before fabrication begins, instead the project embraces a more dynamic, semi-autonomous method. Additive manufacturing is used to create initial volumes in clay, which are deliberately left incomplete. Subsequently, subtractive processes are introduced, allowing the robot to sculpt and refine the surface based on ongoing sensing and perception (Fig. 5.53). The robot's

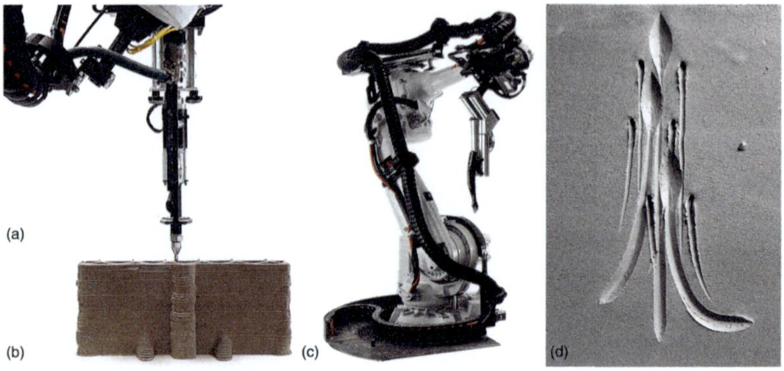

Figure 5.53 Overview of robotic fabrication workflow: (a, b) additive and subtractive manufacturing process; (c) custom robot end-of-arm tool; (d) carving effects produced through directional gestures. Courtesy of Robert Stuart-Smith.

sensors observe the current state of the clay, its depth, texture, and deviations from the intended geometry, and provide real-time data that is used to guide the robot's movements. A custom-designed end-of-arm tool, equipped with a camera and laser sensor, gathers three-dimensional information about the clay's surface and compares it to the target model (Fig. 5.54). Differences between the current and intended forms are translated into vector fields, which then generate custom carving trajectories that evolve over time. The paths are sent continuously to the robot, enabling it to respond to the material and make necessary changes during the fabrication process.

Conventional milling relies on tools that function as neutral cutters, designed for precision and uniformity. In contrast, the robot actuator developed for this project plays an expressive role. A series of physical tests informed the shaping of the tool's geometry, which was carefully refined to produce varied marks, scooping, slicing, pushing, and smoothing the clay in multiple directions. As the tool engages with the material, it leaves behind patterns shaped by orientation, speed, and depth

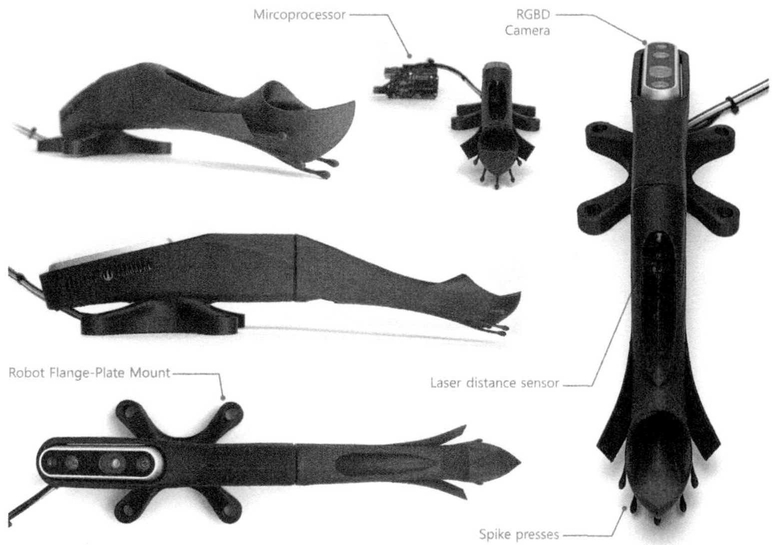

Figure 5.54 Custom robot end-of-arm tool profiled for sculpting, equipped with an RGB-D camera and a laser distance sensor. Courtesy of Robert Stuart-Smith.

of movement, and these gestures are not eliminated in post-processing but intentionally preserved as an integral part of the architectural expression. The resulting surface serves as a continuous record of motion and perception, shaped through an ongoing exchange between sensing systems, tool behavior, and the resistance of the clay, creating a story of its own and allowing the project to revisit the role of ornament in architecture. *Viscera/L* does not apply ornament as a separate layer of meaning or decoration; instead, it generates ornament through the embodied act of fabrication; where structural and ornamental intentions are integrated within a single process. The result does not follow traditional semantic conventions but emerges from the way the robot carves, adjusts, and responds to variations in material behavior (Figs. 5.55 and 5.56). As in Baroque surfaces, where small, partial motions

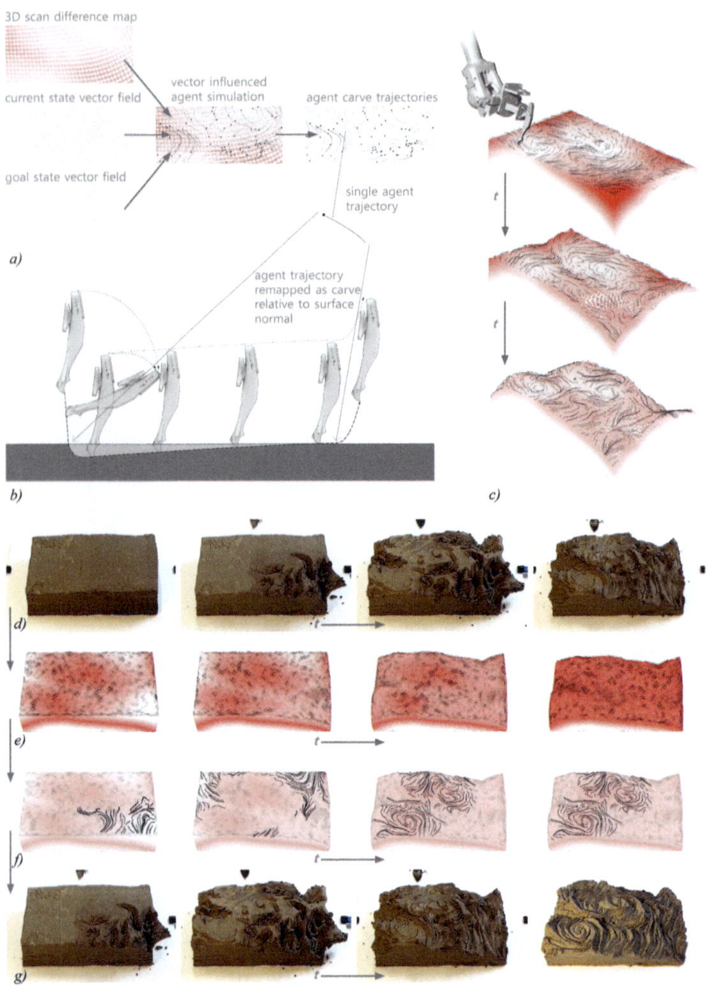

Figure 5.55 Autonomous carving sequence. Top to bottom: initial clay slab state, depth scanning stage, vector field analysis, planned carving trajectories, and final sculpted result. Courtesy of Robert Stuart-Smith.

build into larger visual effects, the robotic trajectories in *Viscera/L* produce complex textures that suggest directionality and depth. These textures adapt to local conditions such as recesses, ridges, transitional zones and create a differentiated surface that stays grounded in the underlying geometry.

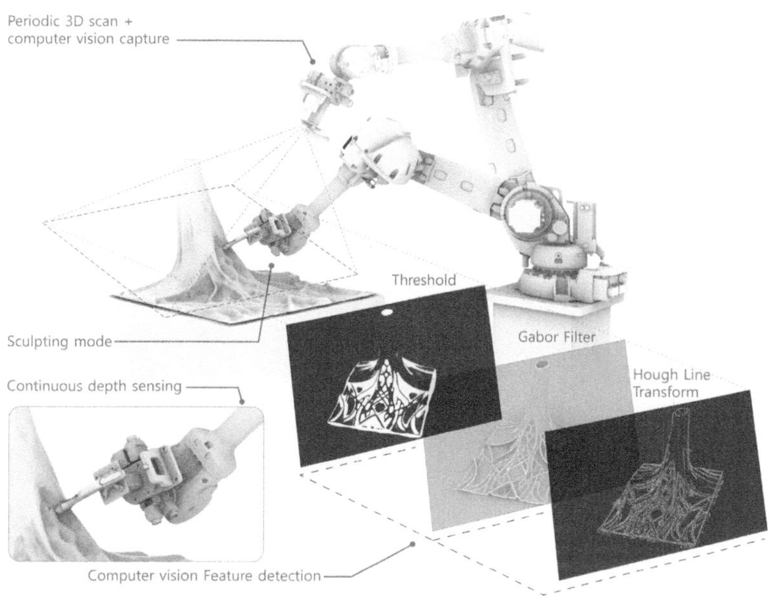

Figure 5.56 Autonomous perception integrates continuous laser sensor data with periodic RGB-D scanning and computer vision feedback throughout the manufacturing process. Courtesy of Robert Stuart-Smith.

The interplay between additive and subtractive methods is most evident in the fabrication of ceiling elements (Fig. 5.57) where transitions exist between columns and overhead surfaces (Fig. 5.58). In one example, a clay slab is printed with overhangs and recessed pockets that would typically be difficult to achieve through extrusion alone. After the initial print, the robot returns to these complex areas to sculpt them further, removing temporary supports and refining edges. In other cases, the process begins with a thick mass of clay shaped more loosely, and the final form is gradually uncovered through carving. In both approaches, the goal is not to produce a flawless surface, but to allow the physical process to leave visible traces, marks that reflect how the material was shaped, adjusted, and resolved over time (Fig. 5.59). These operations highlight the temporal and uncertain nature of fabrication and communicate that

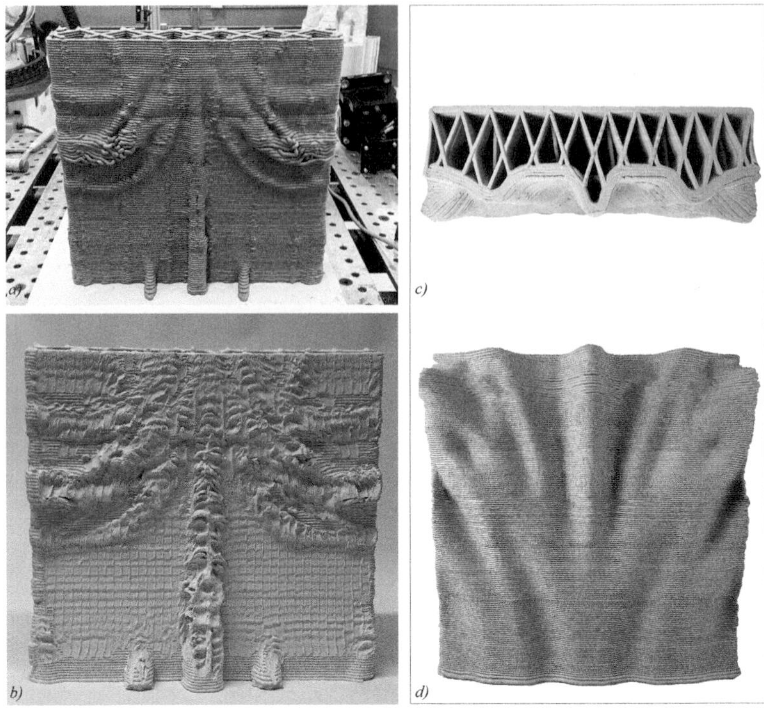

Figure 5.57 Additive and subtractive manufacturing of a ceiling chunk prototype. Robotic sculpting trajectories are generated in response to 3D scan and depth sensor data. Courtesy of Robert Stuart-Smith.

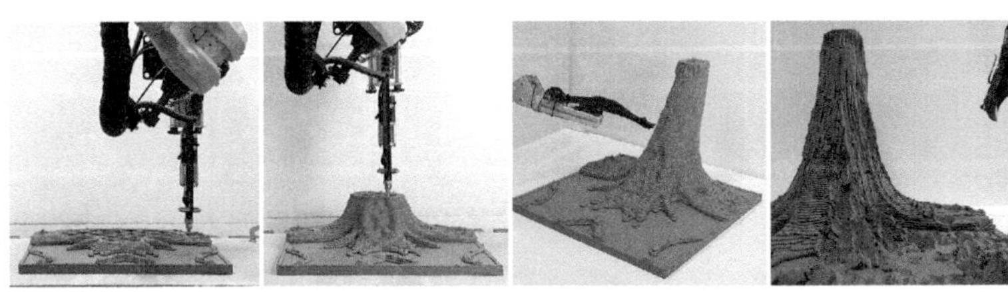

Figure 5.58 Left to right: additive manufacturing of an initial column-to-ceiling transition test, followed by a sculpting prototype. Courtesy of Robert Stuart-Smith.

even when the process begins from the same digital inputs, no two outcomes are identical (Fig. 5.60). Variations in initial volume or surface conditions result in visibly distinct forms where the differences serve as evidence of how form evolves over time. With each

Figure 5.59 Ceiling chunk prototype close-up. The additive and subtractive manufacturing approach seeks to embody character through developed modes of semi-autonomous perception and robotic action. Courtesy of Robert Stuart-Smith.

pass, sensor feedback and tool motion accumulate, leaving behind a layered record of decisions and adjustments. Real-time scanning and high-frequency control help maintain consistency within an intentionally narrow range, while still allowing for the unpredictability that links robotic making to traditions of craft.

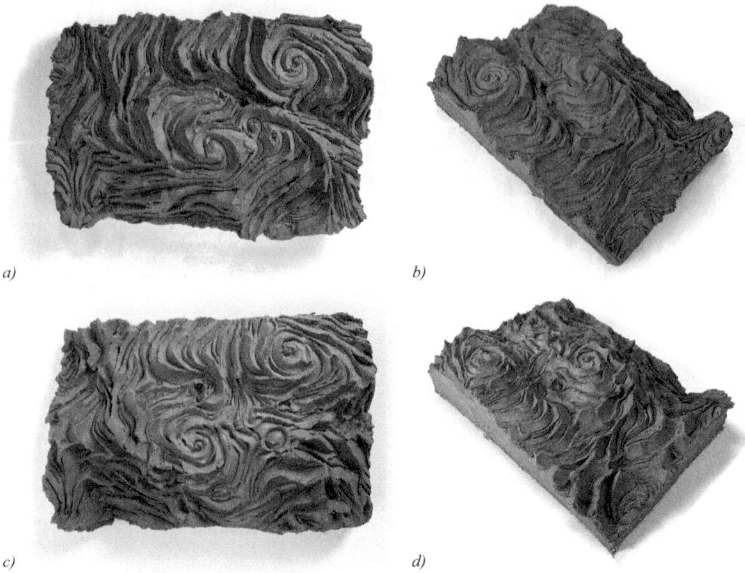

Figure 5.60 Two clay slab volumes are autonomously sculpted into a shared target form. Though the intended geometry remains consistent, subtle differences in robotic carving operations produce distinct surface patterns and trajectories across each version. Courtesy of Robert Stuart-Smith.

Viscera/L places human designers within a broader field of interaction that includes tools, machines, materials, and code. Design decisions unfold not only within the digital model but also through physical testing, calibration, and observation. The designer defines parameters but also remains attuned, responding to how subtle shifts in the clay or changes in tool behavior affect the whole. The project also reflects a post-digital interest in object-to-object relations and non-human perception, where autonomous systems visually assess material conditions and translate that information into immediate, physical responses. These forms of machine vision and feedback blur the line between sensing and making, allowing robots to operate not only as tools but as participants in a shared process of formation. In doing so, the work points toward new models of architectural production, exploring how adaptive systems can generate variation and contribute to a unique element. Automating the bespoke becomes viable only when processes are structured to respond to material behavior. Even when following a shared logic, individual components emerge with subtle differences, and these variations accumulate into a spatial atmosphere that echoes the richness of hand-carved detail found in historical architecture. When robotic actions drift off course or produce unintended outcomes, human input can intervene, either in the moment or retrospectively, to redirect the process, maintaining authorship while still embracing the productive openness of machine agency.

As digital tools continue to evolve, this method opens possibilities for reconnecting contemporary fabrication with the tactile and temporal history of architectural making. The project situates robotic processes within traditions of sculptural detailing and material responsiveness, reimagining not only how buildings are assembled, but how they express meaning. The forms developed in *Viscera/L* do not replicate historical precedents. Yet, in imagining this methodology as a possible approach to the repair of existing conditions, they resonate with a shared sensibility rooted in touch, motion, and the expressive behavior of matter (Stuart-Smith 2024). In doing so, the project draws a line between past and future, grounded in the belief that architecture should be felt as well as seen.

Notes

1 Concept elaborated from a lecture by Karl Daubmann, delivered as part of the PhD lecture series for the International Doctorate in Architecture and Urban Planning at the University of Ferrara, Department of Architecture. November 2019.

2 See "MIT Algorithm Helps Robots Better Predict Human Movement," published in *The Robot Report*, and "Construction Robots Learn to Excavate by Mimicking Humans," featured in *IEEE Spectrum*.

3 The Technology Readiness Level (TRL) scale is a standardized framework used to assess the maturity of a technology, originally developed by NASA and later adopted by institutions such as the European Commission. It ranges from TRL 1 (basic principles observed) to TRL 9 (technology proven in operational environment), offering a structured way to evaluate the development stages of innovations across research, prototyping, and deployment.

4 The Sagrada Familia has inspired experimentation at all scales and in the realm of AI; Daniel Bolojan's "Gaudi + Neural Networks" project uses deep learning (CycleGAN) on Sagrada Familia and natural form datasets to uncover and reinterpret Gaudí's geometric principles as a learning opportunity for design composition.

5 The reconstruction of Palmyra's arch was executed by Giacomo Massari and Filippo Tincolini, with consultancy by Enrico Dini. The project was led by Roger Michel and Alexy Karenowska from the Institute for Digital Archaeology.

6 Centauroos is s startup based in Ascoli Piceno, Italy, and the founding team consists of three architects, Marco Galasso, Giovanni Marinelli, and Roberto Cognoli, with extensive experience in sustainability, urban planning, parametric architecture, and robotics. See: https://www.centauroos.eu/.

7 The project *Traditional House of the Future* has been featured in online publications such as ArchDaily and Dezeen.

8 Malgorzata Zboinska is an Associate Professor in Architectural Design, Digital Technology and New Media Art at the Department of Architecture and Civil Engineering at Chalmers University of Technology in Gothenburg, Sweden. The research presented in section was supported by the Digital Twin Cities Centre funded by Sweden's Innovation Agency Vinnova, grant no. 2019–00041.

9 The research was conducted by Ruxandra Chiujdea, Konrad Sonne, Paul Nicholas, Carl Eppinger, and Mette Ramsgaard Thomsen at the Centre for Information Technology and Architecture (CITA) in Copenhagen, Denmark.

10 The project was conducted by Brandon Clifford, Nazareth Ekmekjian, Patrick Little, and Andrew Manto, and was presented at the 2014 *Robotic Fabrication in Architecture, Art, and Design* conference hosted at the University of Michigan, Taubman College of Architecture and Urban Planning.

11 The research presented in this section was supported by the Swedish Energy Agency Energimyndigheten, grants no. P2022-00865 and P2024-02409.

12 From the Greek word for "fish," the ichthys is an ancient Christian symbol consisting of two intersecting arcs that form the outline of a fish. Early Christians used it as a discreet sign of faith and recognition, and it frequently appears as a decorative or symbolic motif in religious architecture and artifacts.

13 *Viscera/L* invokes a bodily and affective dimension of architectural production, aligning with the project's focus on gestural fabrication, posthuman perception, and richly textured formal language that arises from within the material itself, rather than being imposed from outside.

14 *Viscera/L* was developed by Robert Stuart-Smith, Riley Studebaker, Mingyang Yuan, Nicholas Houser, and Jeffrey Liao at the Autonomous Manufacturing Lab, Weitzman School of Design, University of Pennsylvania.

References

Baird, J. A., Zena Kamash, and Rubina Raja. 2023. "Knowing Palmyra: Mandatory Production of Archaeological Knowledge." *Journal of Social Archaeology* 23 (1): 76–98.

Balzani, Marcello. 2011. *Restauro, Recupero, Riqualificazione. Il Progetto Contemporaneo Nel Contesto Storico*. Skira.

Balzani, Marcello, and Riccardo Dalla Negra. 2017. *Architettura e Preesistenze. Premio Internazionale Domus Restauro e Conservazione Fassa Bortolo*. Skira.

Brandi, Cesare. 1963. *Teoria del Restauro*. Edited by Licia Vlad Borrelli and Joselita Raspi Serra. Ed. di Storia e Letteratura.

Brell-Cokcan, Sigrid, and Johannes Braumann. 2013. *Rob|Arch 2012: Robotic Fabrication in Architecture, Art and Design*. Springer Science & Business Media.

Burry, Mark. 1996. "Parametric Design and the Sagrada Familia." *ARQ: Architectural Research Quarterly* 1 (4): 70–81.

Burry, Mark. 2005. "Homo Faber." *Architectural Design* 75 (4): 30–37.

Burry, Mark. 2012. "Models, Prototypes and Archetypes Fresh Dilemmas Emerging from the 'File to Factory'Era'." In *Manufacturing the Bespoke: Making and Prototyping Architecture*, Edited by Bob Sheil. John Wiley and Sons.

Burry, Mark. 2016. "Robots at the Sagrada Familia Basilica: A Brief History of Robotised Stone-Cutting." In *Robotic Fabrication in Architecture, Art and Design 2016*, Edited by Dagmar Reinhardt, Rob Saunders and Jane Burry. Springer.

Carbonara, Giovanni. 1997. *Avvicinamento al Restauro: Teoria, Storia, Monumenti*. 1st ed. italiana. Liguori.

Carbonara, Giovanni. 2011. *Architettura d'oggi e Restauro. Un Confronto Antico-Nuovo*. Utet.

Chiujdea, Ruxandra, Konrad Sonne, Paul Nicholas, Carl Eppinger, and Mette Ramsgaard Thomsen. 2024. "Design Strategies for Repair of 3D Printed Biocomposite Materials." *Accelerated Design. Proceedings of the 29th CAADRIA Conference*, 311–21.

Clifford, Brandon, Nazareth Ekmekjian, Patrick Little, and Andrew Manto. 2014. "Variable Carving Volume Casting: A Method for Mass-Customized Mold Making." In *Robotic Fabrication in Architecture, Art and Design 2014*, Edited by Wes Mcgee and Monica Ponce de Leon. Springer.

Codarin, Sara. 2016. "Processi Innovativi Di Conservazione e Recupero Del Patrimonio Culturale." *L'Ufficio Tecnico*, no. 7/8 July/August: 12–21.

Dalla Negra, Riccardo. 2011. "Il Restauro Consapevole: la Traduzione dei Principi Conservativi e il Difficile Rapporto con le Preesistenze." In *Il Progetto Contemporaneo nel Contesto Storico*, Edited by Marcello Balzani. Skira.

Dalla Negra, Riccardo. 2017. "Architettura e Preesistenza: Quale Centralità?" In *Architettura e Preesistenze. Premio Internazionale Domus Restauro e Conservazione Fassa Bortolo*. Skira.

De Angelis d'Ossat, Guglielmo. 1995. *Sul Restauro Dei Monumenti Architettonici: Concetti, Operatività, Didattica*. Vol. 13. Strumenti. Bonsignori Editore.

Denker, Ahmet. 2017. "Rebuilding Palmyra Virtually: Recreation of Its Former Glory in Digital Space." *Virtual Archaeology Review* 8 (17): 20–30.

Germanà, M. L., and A. Sposito. 2001. "La Manutenzione Programmata dei Siti Archeologici." *Sposito A. (a Cura Di), Morgantina e Solunto. Analisi e Problemi Conservativi, DPCE, Palermo.*

Kamash, Zena. 2017. "'Postcard to Palmyra': Bringing the Public into Debates over Post-Conflict Reconstruction in the Middle East." *World Archaeology* 49 (5): 608–22.

Kamash, Zena. 2018. *Reviving Palmyra in Multiple Dimensions: Images, Ruins and Cultural Memory*. Taylor & Francis.

Karami, A., F. Menna, and F. Remondino. 2021. "Investigating 3D Reconstruction of Non-Collaborative Surfaces through Photogrammetry and Photometric Stereo." *The International Archives of the Photogrammetry, Remote Sensing and Spatial Information Sciences* 43: 519–26.

McGee, Wes, and Monica Ponce De Leon, eds. 2014. *Robotic Fabrication in Architecture, Art and Design 2014*. Springer International Publishing. https://doi.org/10.1007/978-3-319-04663-1.

Minguzzi, Gianluca. 2008. *Architettura Sostenibile. Una Scelta Responsabile per Uno Sviluppo Equilibrato*. Skira editore.

Oropallo, Gabriele. 2019. "Repair as Prefigurative Politics." *Design Culture: Objects and Approaches*, 157, 157–69.

Ramsgaard Thomsen, Mette, Gabriella Rossi, Anders Egede Daugaard, Arianna Rech, and Paul Nicholas. 2023. "Extending the Circular Design Framework for Bio-Based Materials: Reconsidering Cascading and Agency through the Case of Biopolymer Composites." *World Congress of Architects*, 635–46.

Ratti, Carlo. 2025. *Intelligens, Natural, Artificial, Collective: Biennale Architettura 2025*. Prima edizione. La Biennale di Venezia.

Reinhardt, Dagmar, Rob Saunders, and Jane Burry, eds. 2016. *Robotic Fabrication in Architecture, Art and Design 2016*. Springer International Publishing. https://doi.org/10.1007/978-3-319-26378-6.

Remondino, Fabio, and Sabry El-Hakim. 2006. "Image-based 3D Modelling: A Review." *The Photogrammetric Record* 21 (115): 269–91.

Remondino, Fabio, Sabry F. El-Hakim, Armin Gruen, and Li Zhang. 2008. "Turning Images into 3-D Models." *IEEE Signal Processing Magazine* 25 (4): 55–65.

Rossi, Gabriella, Ruxandra-Stefania Chiujdea, Laura Hochegger, et al. 2023. "Integrated Design Strategies for Multiscalar Biopolymer Robotic 3d Printing." *ACADIA 2022: Hybrid & Haeccetties*, 346–55.

Rudin, R., M. Zboinska, S. Sämfors, and P. Gatenholm. 2022. "RePrint: Digital Workflow for Aesthetic Retrofitting of Deteriorated Architectural Elements with New Biomaterial Finishes." *Proceedings of the 42nd Annual Conference of the Association of Computer Aided Design in Architecture (ACADIA)*. 336–45.

Ruskin, John. 1900. *The Nature of Gothic: A Chapter from the Stones of Venice*. Allen.

Stuart-Smith, Robert. 2024. *Behavioural Production: Semi-Autonomous Approaches to Architectural Design, Robotic Fabrication and Collective Robotic Construction*. Routledge.

Stuart-Smith, Robert, Riley Studebaker, Mingyang Yuan, Nicholas Houser, and Jeffrey Liao. 2022. "Viscera/L: Speculations on an Embodied, Additive and

Subtractive Manufactured Architecture." *Traits of Postdigital Neobaroque: Pre-Proceedings (PDNB)*, Edited by Marjan Colletti and Laura Winterberg. Innsbruck: Universitat Innsbruck.

Thomsen, Mette Ramsgaard, Phil Ayres, Bob Sheil, and Marilena Skavara. 2024. "Introduction: Fabricate 2024: Creating Resourceful Futures." In *Fabricate 2024*. UCL Press.

Thomsen, Mette Ramsgaard, and Martin Tamke. 2022. "Towards a Transformational Eco-Metabolistic Bio-Based Design Framework in Architecture." *Bioinspiration & Biomimetics* 17 (4): 045005.

Willmann, Jan, Philippe Block, Marco Hutter, Kendra Byrne, and Tim Schork, eds. 2019. *Robotic Fabrication in Architecture, Art and Design 2018: Foreword by Sigrid Brell-Çokcan and Johannes Braumann, Association for Robots in Architecture*. Springer International Publishing. https://doi.org/10.1007/978-3-319-92294-2.

Wölfflin, Heinrich, Guy Ballangé, and Bernard Teyssèdre. 1967. *Renaissance and Baroque*. Cornell University Press Ithaca.

Zaffagnini, Mario. 1981. *Progettare nel Processo Edilizio*. Luigi Parma.

Zboinska, Malgorzata A. 2021. "Architectural Research in Hybrid Mode: Combining Diverse Methods within Design-Based Architectural Research Inquiry." *ARENA Journal of Architectural Research* 6 (1).

Zboinska, Malgorzata A. 2024. "Biobased Coatings for Architectural Timber Applied Using the Robotic 3D Printing Technique." *Architectural Science Review*, November 13, 1–18. https://doi.org/10.1080/00038628.2024.2427711.

Zboinska, Malgorzata A., and Delia Dumitrescu. 2021. "On the Aesthetic Significance of Imprecision in Computational Design: Exploring Expressive Features of Imprecision in Four Digital Fabrication Approaches." *International Journal of Architectural Computing* 19 (3): 250–72.

Zboinska, Malgorzata A., and Frederik Göbel. 2025. "Digital Tool Integrations for Architectural Reuse of Salvaged Building Materials." *Automation in Construction* 170 (February): 105947. https://doi.org/10.1016/j.autcon.2024.105947.

Zboinska, Malgorzata A., Isac Mjörnell, Sebastian Oguz, Rebecka Rudin, and Toste Skånberg Dahlstedt. 2022. "Non-Standard Robotic 3D Printing for Architects – A Comprehensive Digital Fabrication Lab Pedagogy Integrating Non-Programmable Material Effects." 19–28. https://doi.org/10.52842/conf.ecaade.2022.1.019.

Chapter 6

Design Agency in the Fourth Industrial Revolution

DOI: 10.4324/9781003590873-6

The Fragmentation of the Traditional Building Process

The assumption that technology alone can resolve inefficiencies on the construction site is misleading. Simply updating machinery or adopting digital tools does not equate to cultural or architectural progress. Technology is an instrument, not a solution, and securing its relevance in architectural production requires multiple forms of intelligence, ranging from material knowledge to fluency in procedure and process. The real challenge lies not in technological advancement, but in its convergence with architecture and restoration practices that utilize tools not yet inherently situated, and requiring recontextualization. Robotic arms, for instance, are generic machines designed for adaptability, while artificial intelligence is currently being developed in a vacuum, detached from specific cultural, environmental, or regulatory context. Prefabrication tends to occur off-site, where it can be optimized for repetition and speed, whereas architecture is by nature on-site, weathered, unique, and rooted in local constraints that differ geographically. So how can we align technological and environmental forces in a meaningful way within our built environment?

The traditional building process is a sequence of complex operations performed to transform a system (Argiolas et al. 2013) and involves the coordination of materials, labor, and resources to produce a physical structure shaped by site-specific conditions and time constraints. In this view, traditional construction follows organized decision-making phases (Briggs 1925) and operational strategies, requiring the investment of limited resources to produce a "superior unity for each phase of transformation" (Ossola and Jona 1999). Long before digital fabrication reshaped contemporary design thinking, architecture had begun to move away from a linear and static conceptual approach. This early shift pointed toward a more dynamic and operational understanding of design, supporting new relationships between intention, process, and tooling. Decision-making phases continue to play a critical role in how construction is organized, where on-site logistics need clear processes and direction to ensure that resources, timelines, and responsibilities align, especially when coordinating the efforts of several trades at once. This kind of sequential thinking traces back to artisan practices, where designers worked in close coordination with skilled fabricators and maintained direct control over materials and technologies. Traditional craft production makes evident this type of consecutive logic, which is structured around a step-by-step process completing each phase before the next can begin. In fields like metallurgy, for instance, key operations are defined by distinct thresholds that separate one task from the next (Zaffagnini 1981).

Industrialization, however, weakened the connection between design and construction. Production became delocalized, and the simplification of technological elements enabled mass replication. This shift "has determined a flattening of the artisan interpretation" (Spadolini 1981) and introduced design constraints driven by downstream production processes, forcing architects to work within the limits of prefabricated components and standard systems (Groak 2002). Nevertheless, throughout the 20th century, evolving approaches to prefabrication opened new possibilities and architects began to explore how variation could be introduced within mass production. Conceptual groundwork for mass customization was initiated—a model never fully realized, but defined as the ability "to produce differentiated elements with the same economic efficiency with which the standardized ones are produced" (Davis 1997). This idea emerged in response to the computerization of manufacturing processes during the Third Industrial Revolution and was further reinforced by the growing accessibility of digital tools (Pine and Davis 1993). The 1980s ushered in what has been described as the Digital Turn, a period in which design began to shift from hand-making to mechanical and eventually digital making (Carpo 2017). Through algorithmic design and digital manufacturing, designers embraced the concept of "multiple variations" (Deleuze 1993) and the ability to generate form through code. Even alongside this progress, architecture struggled more than other sectors to assimilate these changes.

Buildings continue to be treated as unique prototypes (David 1989), "an almost experimental product that responds to the creative ability of the architect and the workers" (Spadolini 1981). Each construction project responds to a "temporary coalition of people and organizations" (Groák 1994), and even with the introduction of advanced tools, the process remains rooted in manual control, visual feedback, and imprecise positioning (Balaguer 2000). This condition of uniqueness, where every project unfolds on a specific site under singular constraints, becomes even more pronounced in the realm of architectural conservation, where historical and cultural specificity both play a central role (Bon 1991).

On-site customization, no longer only the domain of master builders, can now be driven by advanced machines and generative tools that do not mechanize tradition, but build upon it. Challenges continue to persist in scaling digital workflows from fab lab environments to construction sites because of regulatory and safety constraints, but broader shifts are in motion. New technological developments present opportunities to reduce the longstanding fragmentation of the building site, where disconnected trades and processes have historically limited coordination and slowed production. However, a critical gap remains because many of these technologies still operate independently of site-specific knowledge. Bridging these areas requires more than technical proficiency; it demands the encoding of human intelligence into the processes themselves. This means embedding tacit knowledge, material sensibilities, and spatial reasoning into the tools now shaping the future of architecture. Technological advancement, ecological urgency, and cultural preservation are converging into the perfect storm where the task at hand becomes clear. The goal is not to replicate the past, but to update practices, extending care, adaptability, and intelligence into an architecture that has always required advanced tools meaningfully situated in their context.

Applications, Prerogatives, and Materials

Across its history, restoration has evolved from emphasizing stylistic revival to focusing on more nuanced understandings of authenticity, integrity, and historical layering. From Viollet-le-Duc's vision of idealized completion to Ruskin's rejection of intervention altogether, the field has continuously been shaped by competing prerogatives. This plurality of approaches also characterizes the Italian theoretical tradition, where much effort has gone into establishing clear definitions and asking the fundamental question of what qualifies as a historic architectural work, or what it means to carry out a "legitimate transformation" (Carbonara 1997). In this regard, Roberto Masiero, architect and professor of architectural history, outlines the principal positions advanced by major Italian restoration theorists in the chapter *Nel Definire il Restauro, Defining Restoration*. Beginning with the very definition of an *object* of restoration, the perspectives vary widely. Amedeo Bellini defines it as *the pre-esistence*; Giovanni Carbonara says it is *a work of historical interest*; for Stella Casiello it's seen as *the traces of the past;* Roberto Cecchi, defines it as a *heritage asset*; Marco Dezzi Bardeschi believes it to be a *material good inherited from history*; for Paolo Fancelli, it is *what is deemed significant from the past*; Paolo Marconi identifies it as *architecture or the urban context*; Gianfranco Spagnesi, characterizes it as *the existing physical space*; and Paolo Torsello, explains it simply as *the work* (Masiero 2005). These varied positions reflect differing concepts of what restoration entails, but most converge around the value of two core principles: intervention and design. Intervention refers to a direct and active engagement with historic material, while design describes the conceptual and analytical process, grounded in a study of the conditions that make the intervention feasible and meaningful. Tools and knowledge, along with historical and aesthetic considerations, must come together to support a creative direction grounded in clarity, context, and intention.

From the object itself to the act of restoration, one proposed interpretation is that restoration is a system of knowledge and techniques that are aimed at safeguarding a work as a source of culture, which must be enhanced using all the means at our disposal (Torsello 2005). The examples of robotic applications introduced in the Chapter 6 resonate with this lineage, extending into new territories of practice. Contemporary digital-physical workflows suggest that restoration does not need to be a fixed methodological position, but can dynamically function as a process that incorporates technological mediation, material responsiveness, and iterative modes of engagement. The introduction of robotic and computational tools brings an emphasis in restoration to shift toward procedural strategies. Design becomes embedded in the sequencing of actions, the logic of toolpath generation, and in the ability to respond to geometric and material complexity. These changes expand the notion of authorship in restoration, prompting new questions about control, delegation, and the evolving role of a designer to shape both process and outcome.

Restoration has traditionally relied on visual memory, artisanal skill, and formal coherence. In contemporary contexts, these capacities are increasingly supplemented by digital scans, parametric modeling, and material simulation. The prerogatives that guide intervention appear to expand, building upon historical and aesthetic concerns as defined by Cesare Brandi, and incorporating modern computational, environmental, and performative considerations as well. The changing context invites a dialogue between emerging techniques and established theories of continuity, differentiation, and reversibility, where each contributes distinct insights to expand restoration as a situated, creative act.

Materials remain central to this discussion. While past craftspeople selected materials for their symbolic meaning or figurative/structural compatibility, today's designers navigate a broader set of parameters, including physical performance, ecological impact, and machine tolerances. In contemporary research focused on robotic fabrication for innovative restoration applications, case studies illustrate that materials such as clay, stone, biopolymers, and recycled aggregates are chosen for the design opportunities they enable, their environmental responsiveness, and their behavior under robotic control. The act of repair becomes both a technical process and a conceptual gesture that responds to loss while constructing a new expression of form.

Using robotic arms as extensions of the human body within restoration has resulted in various capacities, such as the ability to engage fragmented conditions with precision. Yet, as demonstrated in the case studies, many contemporary practices turn to robotic fabrication and computation not to perfect, but ironically to explore imprecision. These approaches embrace digital glitches—unique moments of variation or unpredictability—as a means of revealing the "expressive potential of imperfection" (Zboinska and Dumitrescu 2021). This orientation tends to humanize the process and the tools themselves, softening the boundary between human intention and machine execution. The result is a more intimate engagement with making and repair, especially when working with figurative forms like ornamentation, symbolic motifs, and sculptural details.

Choice of the Right Architectural Language

The case studies examined throughout this book reflect a spectrum of architectural language and reveal how each formal strategy, whether discreet, analogical, or openly contemporary, expresses a specific perspective in regard to working with the past. In any restoration process, the selection of materials and tools must respond to the intentional expressive or linguistic outcome. Whether the aim is to maintain visual continuity or to introduce deliberate contrast, this conceptual stance shapes how the intervention engages with its context on a case-by-case basis, and how its presence is ultimately perceived.

This spectrum is illustrated across the robotically enabled projects presented in the previous chapter. Interventions range from subtle material substitutions and textural blending to more assertive responses defined by geometric contrast or parametric fluidity. Some proposals adopt restrained gestures that quietly dissolve into the existing fabric (*Sagrada Familia*), while others intentionally emphasize the fracture line (*The Traditional House of the Future*), using digital tools to construct expressive forms that mark their own presence through formal autonomy (*Viscera/L*). In each case, new elements may connect to the existing structure through a morphological method or through a clear contrast in form. "Architectural objects bear the scars of the technique by which it was made," writes Tim Ingold, highlighting the inherent expressiveness embedded in making (Ingold 2013). Digital fabrication extends this logic, not by replicating traditional craft, but by digging deeper and revealing the computational structures that guide its processes while opening a space for dialogue between tool and outcome (Bonwetsch et al. 2010).

Legibility can emerge through contrasts in material, chromatic nuance, or spatial configuration. Techniques such as *rigatino*, a method using fine, parallel lines of pigment to complete damaged surfaces without fully concealing the repair, allow for a visual continuity that also acknowledges absence. In more abstract cases, broad tonal field—unified areas of color or tone—or simplified geometries are introduced to signal contrast yet remain respectful of the original.

Computational design expands these expressive strategies and provides a new found freedom to explore. Free-form geometries, unconstrained by orthogonal systems, can generate counter-structures that coexist with historic layers without mimicking them. This layered approach embraces continuity and transformation, but still proposes an adaptive formal vocabulary for restoration that frames the act of repair as both a computer-aided spatial proposition and a cultural interpretation in tandem.

A Note on the Reintegration of Lacunae

A central component of the investigation presented in this manuscript was a lab-based phase to test concepts firsthand. This lab work put into practice the phase of "pushing buttons in the lab" previously introduced in Chapter 1. The outcome was a low-tech, low-risk, low-budget proof of concept that explored robotic clay extrusion to repair a wall subjected to material degradation within a historical building. This experiment provided direct insight into design workflows, tools, and materials, clarifying how they inform robotically augmented restoration decision-making.

The element of focus was a wall section showing material loss, defined in restoration practice as a *lacuna*. The *lacuna* is a recurring theme in the discourse of architectural restoration (Dalla Negra and Varagnoli 2017), particularly when speculating how to best resolve, or reintegrate, material gaps in the fabric of a building. This challenge often serves as a response to the aesthetic of ruin or the allure of the unfinished. In contrast, the approach presented aims to reframe the idea of completeness in a constructive way, both literally and metaphorically.

Lawrence Technological University provided the setting for this phase of work where lab activities were designed to build a clearer understanding of tooling, simulation, and material behavior in real-world conditions. As a test-bed, the investigation focused on the Woodward Avenue Presbyterian Church in Detroit (Fig. 6.1). Built in 1911, this neo-Gothic structure is currently in a state of neglect after years of abandonment. The church holds both architectural and cultural significance as a registered national landmark and is closely tied to Detroit's broader urban trajectory and its cycles of economic prosperity and decline. The church's condition exemplifies a larger pattern across the city, where historically important buildings have been left to deteriorate over the years.

Figure 6.1 Interior view of the Woodward Avenue Presbyterian Church in Detroit. Photo by Sara Codarin.

Initial research focused on data collection using digital surveying tools (Fig. 6.2), including terrestrial laser scanning, handheld photogrammetry, and drone imaging. These methods produced a high-resolution record of existing conditions and areas of degradation. Among several deteriorated elements, a damaged masonry wall was selected as the case study

Figure 6.2 Aerial data capture using drone-based photogrammetry to generate high-resolution surface textures, supporting digital modeling and material analysis. Photo by Sara Codarin.

Figure 6.3 Analysis of internal and external material degradation. Top left: *lacuna* on the masonry wall beneath the dome of the church. Photos by Sara Codarin.

(Fig. 6.3). This non-structural void featured complex geometry, undercuts, and restricted access, presenting a relevant condition for testing robotic repair. The *lacuna* served both as a material challenge and as a conceptual test site, raising questions about how to reintegrate missing material in a way that acknowledges historical absence while proposing a contemporary intervention. A conservation expert might advocate for a *cuci-scuci* technique to restore the formal continuity of that part of the building. Instead, this damaged condition was approached as a design prompt (Fig. 6.4), an opportunity to position technology within the restoration process through experimentation as opposed to an immediate problem-solving workflow.

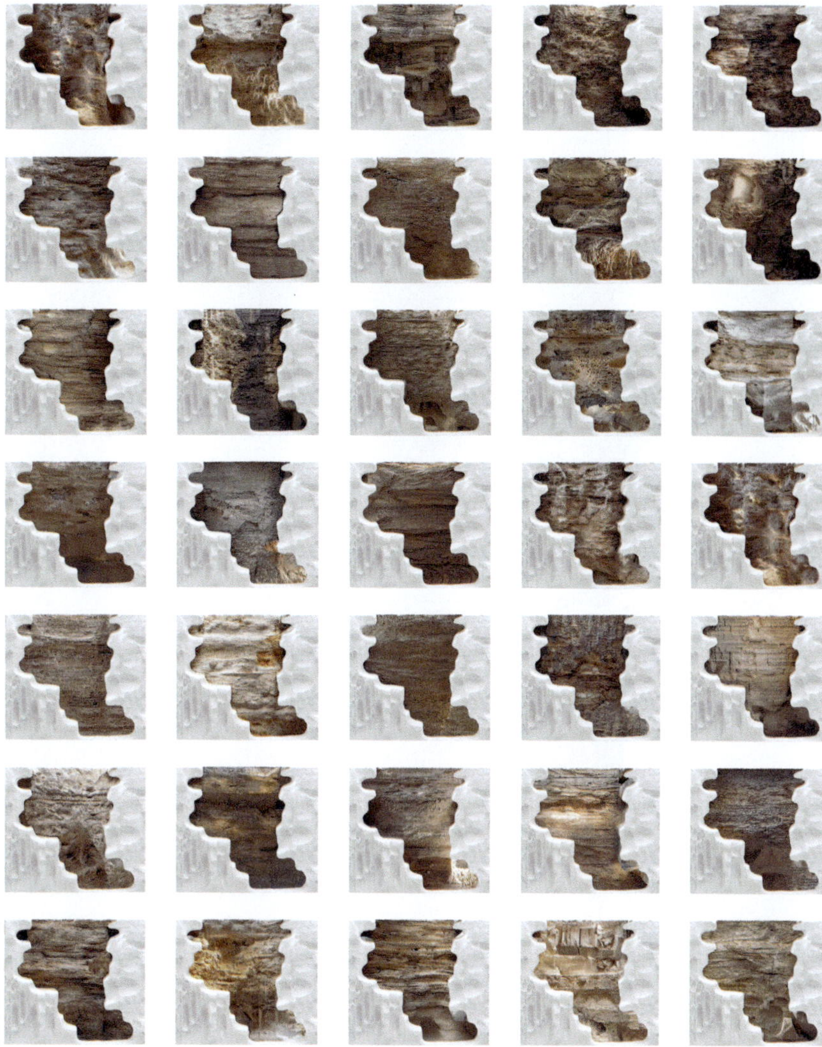

Figure 6.4 Conceptual design exploration for reintegrating material loss in the masonry wall, treating the damage as a design prompt. Visual study created from collaged textures photographed on site.

Survey data were used to develop a digital model of the damaged area, followed by a full-scale mock-up that was fabricated using CNC-milled foam. The mock-up defined a low-resolution but spatially accurate platform to test robotic kinematics, material flow, and toolpath generation. Clay was chosen as the extrusion material due to its low cost, reversibility, and ease of manipulation. A custom extruder end-effector was developed to work in conjunction with a six-axis industrial robotic arm, enabling controlled deposition of clay across irregular geometries (Fig. 6.5). Digital simulations

Figure 6.5 Wall gap fabricated through robotic clay extrusion at 1:1 scale, with a proposed hypothesis for actuators designed to interact with irregular surfaces. Photos by Sara Codarin.

refined toolpaths and predicted potential collisions or kinematic limitations, in turn informing the physical implementation where material was gradually deposited to reconstruct the missing volume. The responsiveness of the material, the reach of the robotic arm, and the performance of the extruder were continuously assessed throughout the process.

This experiment provided a moment of realization—when design began to shift from traditional geometric form-making to code-driven processes (Fig. 6.6). Attention moved away from static representations and toward the logic of toolpaths where the kinematic behavior of the robotic arm took precedence. What emerged was a choreography of motion in space, shaped by the constraints of the task and the responsiveness of the system supporting it. The design process did not necessarily stop at representation but continued through direct engagement with the physical tools of making, where the robot functioned both as a medium and an active participant. The process unfolded through recursive testing and each step was informed by fabrication awareness, suggesting that custom workflows are indeed an essential point of departure when working with emerging tools.

This workflow can be read as a proof of concept, aligned with what in market-based frameworks is referred to as a minimum viable product

Figure 6.6 Typical Grasshopper definition employing the KUKA|PRC plug-in to program an industrial robotic arm.

(MVP). An MVP is defined as a product with just enough features to generate early feedback and guide future development. This language is common in the context of startups, where the idea of an MVP frames experimentation acting as a structured tool for iterative improvement. According to Eric Ries, a MVP is "the fastest way to get through the build-measure-learn feedback loop" (Ries 2011). In this research context, the MVP was not a finished solution but a functional prototype, a test intended to explore the hypotheses itself instead of resolving a fully formed technical question (Fig. 6.7). The experiment aimed to determine

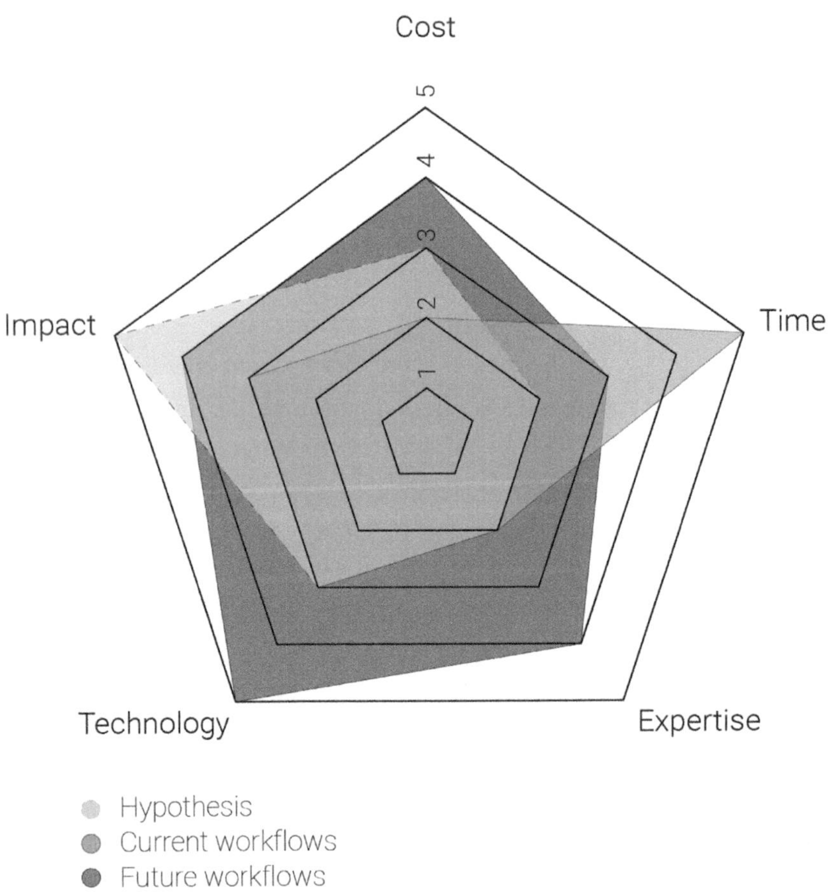

Figure 6.7 Metric for evaluating the relationship between hypothetical scenarios, current technological possibilities, and future workflows.

whether such a process is worth further pursuit, and whether it holds potential for replication, adaptation, or deployment in other contexts. As a prototype, the lab experiment marked a modest but exploratory step toward broader applications in future investigations.

The Seven Points of a Fictional Charter for Robotic Restoration

In light of emerging tools, distributed intelligence, and evolving cultural responsibilities, it is possible to expand upon the foundational ideas of restoration, using historical charters as a reference point to begin addressing the realities of a technological age. This fictional charter proposes a framework in which human and machine agency coexist, and where robotics and digital fabrication are understood as instruments capable of renewing heritage through precision, care, and situated practice.

Article 1. Restoration is an act of cultural continuity that has the mandate to preserve the historical and material integrity of the built environment while enabling new methods that uphold its meaning, dignity, and presence.

Article 2. The aim of restoration is to safeguard the authenticity of the architectural work while acknowledging that methods of intervention evolve. These methods include digital and automation tools when they act in support of cultural memory, care, and situated practice.

Article 3. Restoration must be grounded in a thorough understanding of the object and its context. Each intervention is to be preceded by a diagnostic phase that integrates computational accuracy and human judgment, ensuring that decisions arise from data, experience, and ethical responsibility.

Article 4. The application of robotics and digital fabrication is acceptable when these instruments are used to enhance material understanding, procedural rigor, and site-specific responses. Tools must be selected not for efficiency alone, but for their capacity to support thoughtful and informed restoration.

Article 5. Digital craft serves the transmission of knowledge, enabling embodied craft knowledge to be translated, extended, and re-encoded through programmable processes.

Article 6. Where traditional knowledge is at risk of disappearance, restoration shall include mechanisms for recording, simulating, and transmitting such expertise through digital means. This ensures continuity of cultural knowledge across generations and media.

Article 7. Restoration recognizes the built environment as part of a larger ecological system. The application of technology must support a symbiotic relationship between the historical structure and its natural context, placing regenerative practices above extractive strategies.

Restoration in an age of automation resists both technocratic reduction and nostalgic retreat. These naïve seven points are not a declaration in the tradition of the pioneers, the militants, or the maestros of restoration theory. They simply introduce, even if timidly, a few reflections on how restoration might respond to emerging tools and shifting responsibilities. The opening of this manuscript frames the use of robotics in restoration as an outrageous attempt to provoke a conversation. And perhaps that's the point. Language shapes practice by guiding how we use new tools, how we act, and how we pay attention to materials, places, and time. Addressing and reflecting upon our own interaction within the process is the most human gift we can provide to a promising collaboration that stands before us.

New Skills for Advanced Workflows

In a rapidly evolving technological ecosystem, tools, platforms, and machines do not operate in isolation, nor do they alone carry inherent value. Their significance emerges when they are purposefully integrated, also understood as a form of curation, to contribute meaningfully to design and construction. This integration is the function of workflows (Fig. 6.8), which are more than simply sequences of tasks. Workflows provide the underlying logic that connects tools to intention and aligns operations with outcomes (Marble 2012).

Figure 6.8 Conceptual workflow envisioning future digital restoration processes, combining on-site robotics with advanced diagnostics, data acquisition, modeling, fabrication, and assembly.

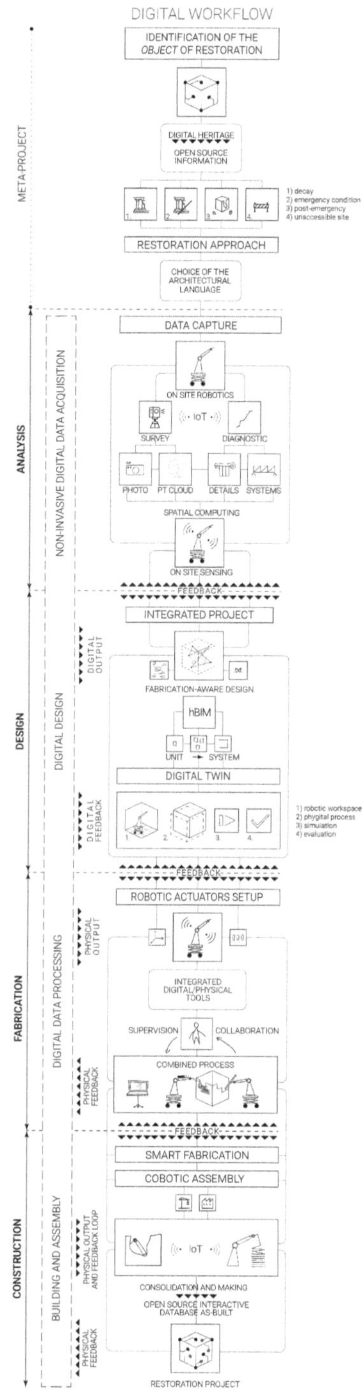

The progression from analysis to design, fabrication, and construction forms the backbone of advanced workflows. Instead of unfolding in a fixed, linear sequence, these phases benefit from being structured as iterative and interconnected loops with the advantage of flexibility because analytical data can inform design decisions in real time. Fabrication constraints can shape modeling logic from the outset. On-site conditions may prompt revisions that ripple back through earlier stages. Pursuing non-linear workflows allows each phase to respond dynamically to the others, supporting a more integrated and intelligent design process.

Within this system, the architect's role expands into new domains. The *digital craftsperson* navigates between code and material behavior, translating geometry into physical output. The *algorithm designer* encodes variability and performance into parametric rules. The *data-informed architect* integrates simulation, environmental inputs, and optimization into early design thinking. These roles converge into a new cohesive figure: the *operator of the Fourth Industrial Revolution* (Fig. 6.9). The operator of the Fourth Industrial Revolution channels the legacy of the medieval master builder, synthesizing collective intelligence shaped by information technology. This figure navigates design and fabrication challenges by coordinating expertise across the team, directing each task to the most capable hands, and basing efficient decisions upon an informed position of proficiency.

To support this shift, architectural education needs to evolve beyond technical instruction. Students must learn to think across scales and systems, connecting abstract ideas with material execution. This includes fluency in computational logic, scripting, systems integration, and an understanding of how materials behave under real-world constraints. The capacity to model, test, and iterate across digital and physical platforms becomes critical. In this way, workflows function not only as organizational tools, but also as pedagogical frameworks, structures that teach students how to navigate and operate within complexity (Fig. 6.10).

Pre-industrial builders learned by trial and error, refining their work through repetition (Carpo 2017). Today, architects operate with digital simulations that allow for continuous iteration before physical production.

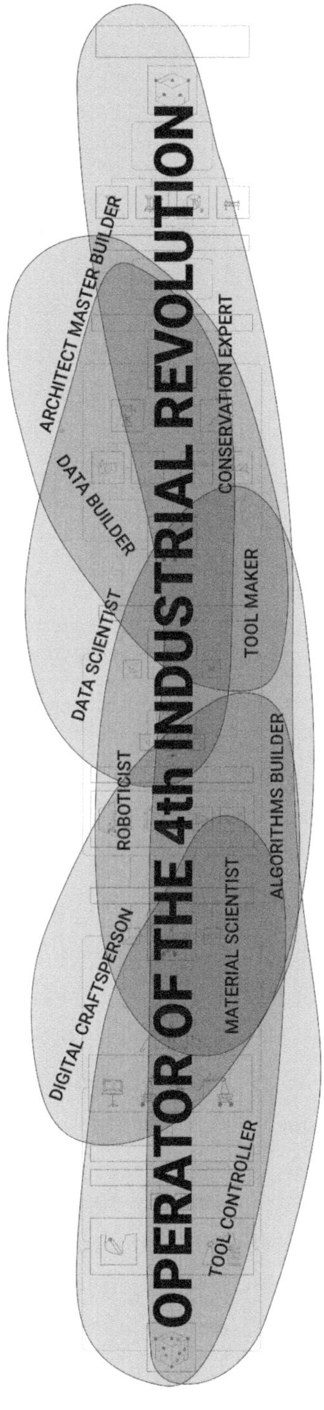

Figure 6.9 Roles within a speculative digital restoration workflow, led by the Operator of the Fourth Industrial Revolution, who orchestrates the work of specialists in diagnostics, surveying, modeling, fabrication, and assembly into an integrated process.

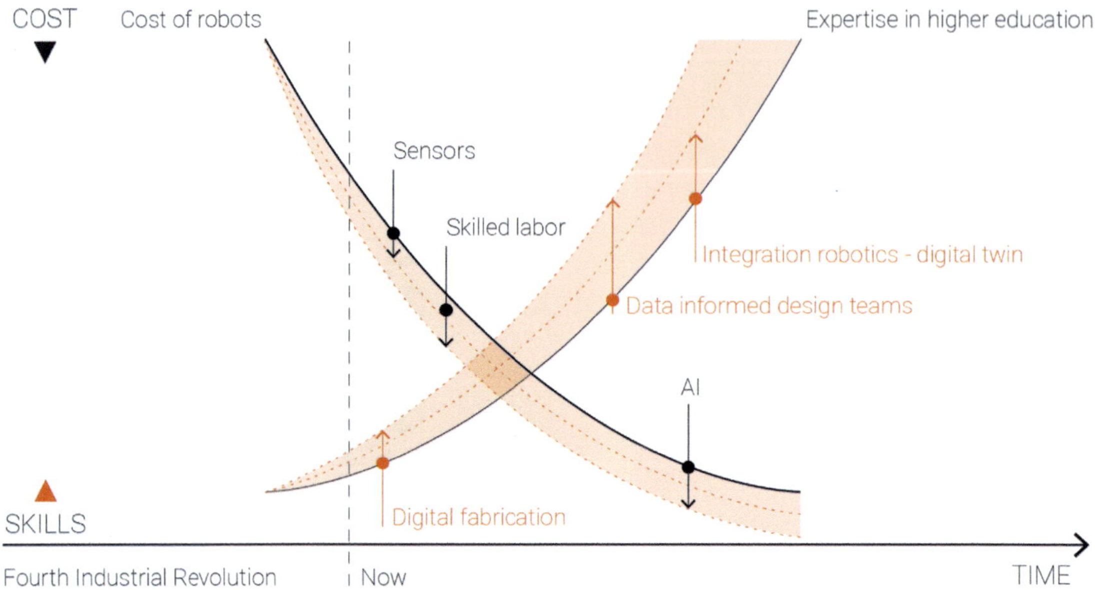

Figure 6.10 The role of higher education in the democratization of robotics within the context of the Fourth Industrial Revolution. A parallel trend of technological knowledge dissemination and decreasing costs of industrial robots, supported by advances in AI and sensor development. A growing gap in skilled labor and rising employment costs as drivers of increased automation. Integration of digital fabrication into academic curricula across scales. Preparation of future professionals for participation in multidisciplinary, data-informed teams.

Design is still a predictive act, but now, it predicts through loops of feedback, data, and analysis. This predictive capacity depends on the strength of the workflow and, therefore, advanced workflows are cultural constructs that shape how architecture responds to emerging technologies. They support both innovation and continuity by aligning intention with execution, creating a new kind of architect—part theorist, part technician, part orchestrator—able to curate the tools and processes that define how we design and make space.

Leveraging the Fourth Industrial Revolution

The Fourth Industrial Revolution introduces tools and processes that are gradually reshaping the construction industry, yet their integration into architectural restoration remains uneven and exploratory. Robotics, automation, and digital workflows continue to advance, but their full application at the infrastructural scale is still unfolding. Restoration, in this evolving landscape, provides a unique space to test and adapt these methods in response to specific material, historical context, and cultural conditions. Current research emphasizes the importance of incremental development, focusing on how digital processes can adjust to the complexities of the built environment through iterative, small-scale applications, avoiding the imposition of fixed technological solutions.

Project methodologies now begin to reflect this shift. Construction can progress through distributed systems where design, fabrication, and assembly unfold in parallel, across multiple sites. Instead of following a sequential timeline shaped by phase-based logic and trade separation, higher-performance workflows make it possible to build elements like foundations and roofs at the same time. These approaches support customized, non-linear processes that respond to logistical complexity and site-specific constraints. Examples from practice, including the Sagrada Familia and BLU Homes, illustrate how components produced in different locations can come together through coordinated sequencing. These scenarios depend on distributed intelligence: teams working across disciplines, informed by shared data and aligned through digital tools. Parametric modeling, spatial data capture, and project management platforms support the integration between design intent and execution, making room for a more continuous and responsive construction logic grounded in ubiquity of access and the potential for autonomy in project delivery (Bieg et al. 2019).

Architecture often pulls from adjacent sectors, such as media, simulation, and product design, to integrate advanced tools into design communication and production (Fig. 6.11). While many of these tools were not originally developed for construction, they gradually shape how buildings are designed and realized. Robotics, in particular, could influence future conservation efforts by enabling precise material interventions, managing complex geometries, and supporting documentation and repair workflows that remain difficult and inefficient to achieve through manual methods alone.

Restoration operates within a framework shaped by historical interpretation, disciplinary theory, and site specificity. Digital tools do not

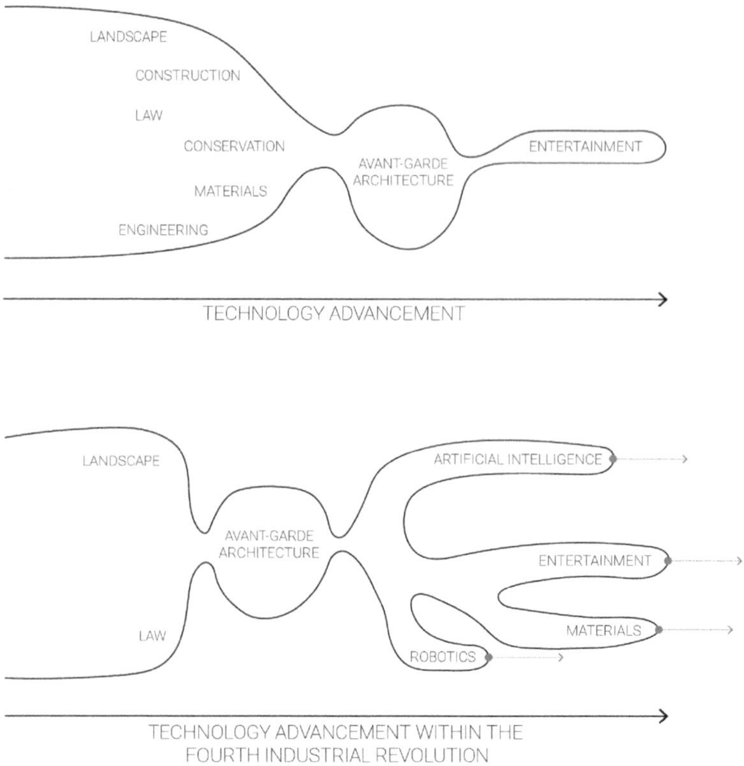

Figure 6.11 Leading sectors in the development of digital technologies. Architecture draws on tools developed in adjacent fields such as entertainment to support design production and project communication. The Fourth Industrial Revolution introduces robotics and automation as key forces shaping technological progress across multiple industries. These developments create conditions for the construction sector to rethink long-standing practices and engage with more adaptive and digitally informed workflows.

replace these structures, but they must be adapted to work within them. As construction continues to absorb emerging technologies, restoration becomes a field where technological knowledge meets humanistic reasoning. This intersection provides the conditions to reframe how innovation is understood and applied, not through efficiency alone, but through theoretical grounding and architectural intent. Research centers and practitioners continue to refine strategies that translate computational processes into situated, intentional design actions. What emerges is a practice that connects physical craftsmanship, digital intelligence, and disciplinary knowledge in ways that reinforce the cultural and spatial complexity of the built environment.

References

Argiolas, Carlo, Giulio Concas, Marco Di Francesco, et al. 2013. "Knowledge in Construction Processes." *KMIS 2013 5th International Conference on Knowledge Management and Information Sharing*, 397–404.

Balaguer, Carlos. 2000. "Open Issues and Future Possibilities in the EU Construction Automation." *Proceedings of the IAARC International Symposium on Robotics and Automation, Taipei, Taiwan*.

Bieg, Kory, Danelle Briscoe, and Clay Odom. 2019. "Ubiquity and Autonomy: Introduction to the Paper and Project Proceedings of the 39th Annual ACADIA Conference." In *Ubiquity and Autonomy: Paper Proceedings of the 39th Annual Conference of the Association for Computer Aided Design in Architecture*, Edited by Kory Bieg, Danelle Briscoe, and Clay Odom. The University of Texas at Austin School of Architecture.

Bon, Ranko. 1991. "What Do We Mean by Building Technology?" *Habitat International* 15 (1–2): 3–26.

Bonwetsch, Tobias, Fabio Gramazio, and Matthias Kohler. 2010. "Digitales Handwerk." In *GAM Architecture Magazine 06*. Springer.

Briggs, Martin Shaw. 1925. *A Short History of the Building Crafts*. The Clarendon Press.

Carbonara, Giovanni. 1997. *Avvicinamento al Restauro: Teoria, Storia, Monumenti*. Liguori Napoli.

Carpo, Mario. 2017. *The Second Digital Turn: Design Beyond Intelligence*. MIT Press.

Dalla Negra, Riccardo, and Claudio Varagnoli. 2017. "Le Lacune Urbane Tra Presente e Futuro." *Atti Della Giornata Di Studi (Pescara, 4 Marzo 2015), GB Editoria, Roma*.

David, T. Yeomans. 1989. *Building Construction before Mechanisation*. JSTOR.

Davis, Stanley M. 1997. *Future Perfect*. Basic Books.

Deleuze, Gilles. 1993. *The Fold: Leibniz and the Baroque*. University of Minnesota Press.

Groák, Steven. 1994. "Is Construction an Industry? Notes towards a Greater Analytic Emphasis on External Linkages." *Construction Management and Economics* 12 (4): 287–93.

Groak, Steven. 2002. *The Idea of Building: Thought and Action in the Design and Production of Buildings*. Taylor & Francis.

Ingold, Tim. 2013. *Making: Anthropology, Archaeology, Art and Architecture*. Routledge.

Marble, Scott. 2012. *Digital Workflows in Architecture: Design–Assembly–Industry*. Walter de Gruyter.

Masiero, Roberto. 2005. "Nel Definire il Restauro." In *Che Cos'è il Restauro? Nove Studiosi a Confronto*, Edited by Paolo Torsello. Marsilio.

Ossola, Francesca, and Simonetta Jona. 1999. *La Gestione Del Processo Edilizio: Pianificazione Progettuale Ed Operativa*. Levrotto & Bella.

Pine, B. Joseph, and Stan Davis. 1993. *Mass Customization: The New Frontier in Business Competition*. Harvard Business School Press.

Ries, Eric. 2011. *The Lean Startup: How Today's Entrepreneurs Use Continuous Innovation to Create Radically Successful Businesses*. Crown Books.

Spadolini, P. 1981. *Progettare nel Processo Edilizio*. Le Monnier, Firenze.

Torsello, Paolo, ed. 2005. *Che Cos'è il Restauro? Nove Studiosi a Confronto*. 1st ed. Elementi. Marsilio.

Zaffagnini, Mario. 1981. *Progettare nel Processo Edilizio*. Luigi Parma.

Zboinska, Malgorzata A., and Delia Dumitrescu. 2021. "On the Aesthetic Significance of Imprecision in Computational Design: Exploring Expressive Features of Imprecision in Four Digital Fabrication Approaches." *International Journal of Architectural Computing* 19 (3): 250–72.

Chapter 7

The Future of Craft, Industry, and Heritage

DOI: 10.4324/9781003590873-7

What Comes Next

Mass customization and its role within architecture verses cultural heritage restoration follow markedly different paths (Fig. 7.1). In contemporary architectural practice, the process begins with stable inputs: client needs, project requirements, and clearly defined performance criteria, enabling designers to develop responses tailored to a specific context. This stability allows the focus to remain on the creation of a specific output over the management of unpredictable starting conditions.

Cultural heritage conservation reverses this logic. Inputs emerge from the existing state of damaged buildings, each shaped by unique patterns of decay, structural instability, and layers of historical transformation. These conditions make the starting point inherently *unstable*, requiring careful, case-by-case analysis before interventions can be planned. The challenge intensifies because of the responsibility to preserve authenticity, maintain historical integrity, and ensure future resilience.

Within this context, producing custom construction components reintegrates a building into its functional and cultural continuum, focusing on maintaining continuity instead of introducing novelty for its own sake.

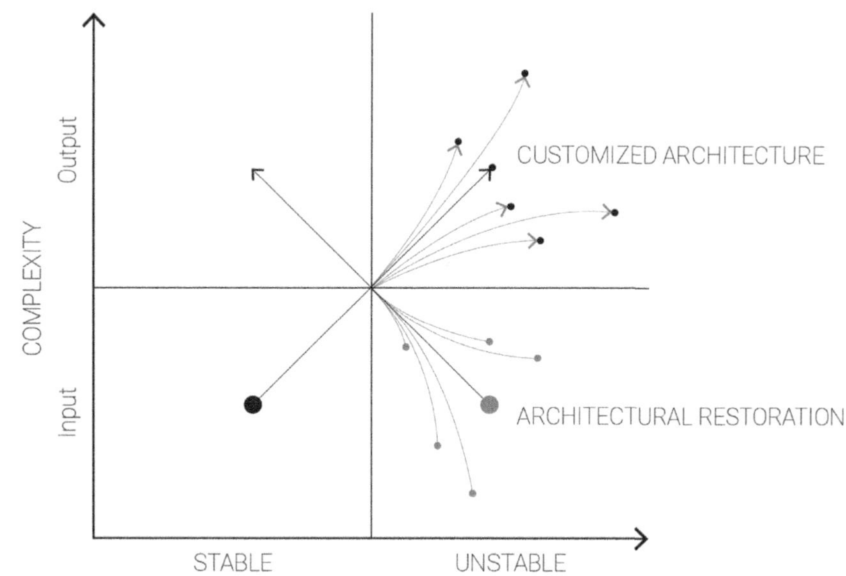

Figure 7.1 Diagram contrasting predictable design scenarios with condition-driven interventions in heritage work, highlighting how varying initial states influence decision-making, fabrication strategies, and the integration of emerging technologies.

Designing site-specific responses bridges contemporary material interfaces with irregular geometries, material inconsistencies, and environmental constraints to safeguard the building's cultural value for future generations.

Advances in computational design and digital fabrication create opportunities so that the process is more systematic. Data capture, parametric modeling, and robotic fabrication can align within workflows that link building diagnosis directly to on-site execution. This speculative and methodological approach remains open to refinement and aims to merge technical precision with a strong theoretical foundation.

It is here that heritage restoration can find a space for technological innovation, and where collaboration between designers, engineers, and machines can enable precise, adaptable, and culturally attuned interventions. This practice has potential to respond dynamically to the complexity of the past while shaping its continuity into the future.

It's Possible

Architectural restoration could fully embrace robotics when the construction sector and digital heritage specialists work in concert with one another. Such alignment would strengthen investment in the field and expand the role of architects, positioning them as central operators within the Fourth Industrial Revolution. This reasoning aligns with the principles of the MacLeamy curve (Fig. 7.2), which shows that the greatest capacity to influence quality and performance occurs early in the design process, when the cost of change is minimal. As projects advance, design flexibility decreases while the expense of modifications escalates. If design-to-fabrication becomes nearly instantaneous, more of the available time can be dedicated to design development itself—testing options, refining details, and improving the quality of the outcome—instead of being consumed by downstream coordination and reactive problem-solving.

In traditional restoration, when additions, integrations, or repair measures are required, many decisions are made directly on site, often with limited digital documentation or predictive planning. This reactive

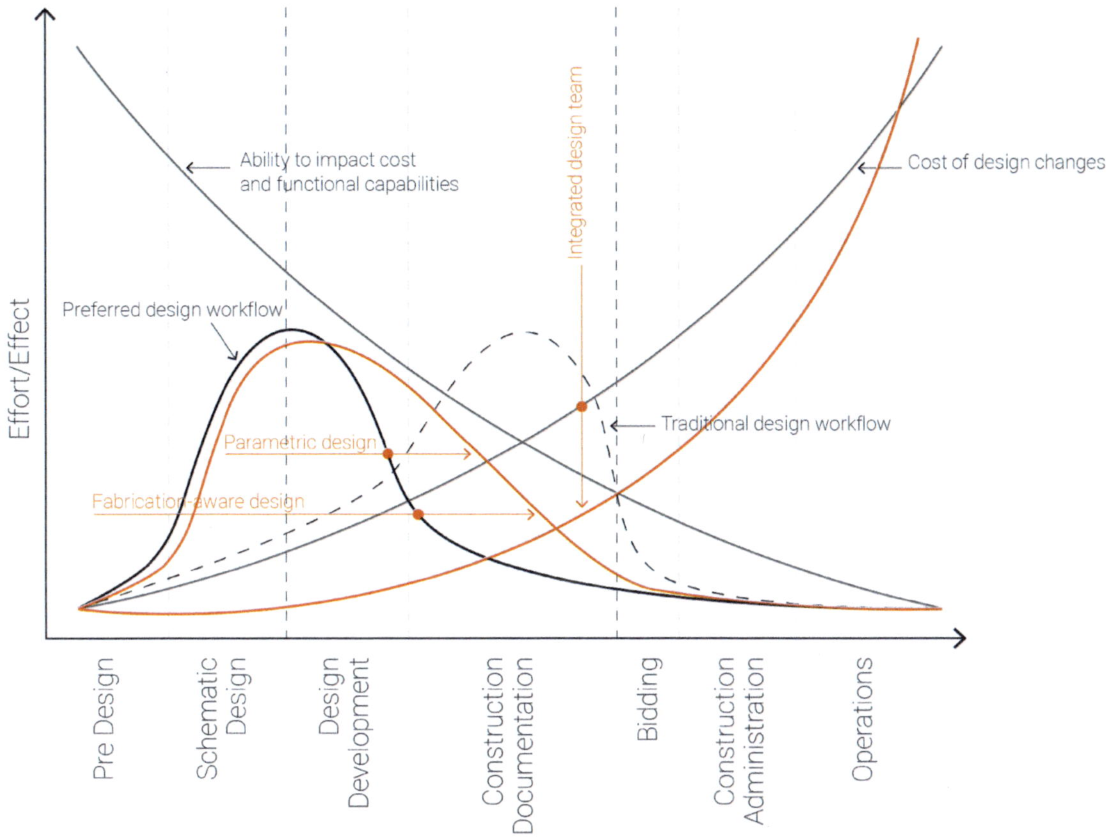

Figure 7.2 *The MacLeamy Curve*, by Patrick MacLeamy. The diagram illustrates that shifting design effort earlier in the process reduces the cost impact of later changes while increasing the influence on the final outcome. Early stage design holds the greatest potential to shape quality at minimal cost. Extending this phase and deepening project development can lead to higher building quality. The Fourth Industrial Revolution enables such a shift by integrating design and production through parametric and fabrication-aware processes.

approach reduces efficiency and makes it difficult to store and transfer knowledge across projects. One reason this transfer is so challenging is that architectural production rarely operates as a repeatable process. Every building is, in effect, a prototype shaped by its materials, site-specific constraints, and design intent. In heritage contexts, this uniqueness is even more pronounced: each structure is a singular expression of culture, informed by historical events, craft traditions, and local conditions. Without a way to systematically capture the specific parameters of each case, much of the knowledge generated in one project cannot be applied directly to another.

A possible way to address this gap is through restoration practices supported by automation and integrated digital workflows. Such practices would rely on tools with user interfaces designed to be solution-oriented, embedding enough contextual information to address specific problems while also encoding human intelligence, in both its embodied and spatial knowledge, into the process. In this view, the act of augmented making can begin to be understood as part of cultural heritage itself, extending into a human–robot–material continuum.

In such a scenario, early stage data—gathered through data capture, diagnostics, and material analysis—could be embedded into parametric and fabrication-aware models, creating a direct link between design intent, technical requirements, and site-specific constraints. These models might then generate machine-readable instructions, allowing robotic systems to engage with the complexities of material interfaces, irregular geometries, fragile surfaces, and assemblies with a level of precision and repeatability that bypass conventional methods.

The technological capacity to work in this way already exists. High-resolution scanning, predictive modeling, and robotic fabrication are available and proven. What remains is the gradual convergence of knowledge from heritage specialists, architects, research centers, and the construction sector—an alignment that would make it possible to embed cultural values and technical objectives from the start. Such a shift would make incremental, positive change viable, while enabling restoration to respond to the specificity of each heritage site and to contribute to a growing body of shared expertise for future interventions globally.

It's Necessary

Broader trends, explained below, indicate that robotics will likely play an increasingly central role in architectural restoration in the years ahead. The influence of early design phases is now widely acknowledged: decisions made at the outset shape both quality and efficiency, and because this stage is underdeveloped in restoration projects, opportunities for precision and foresight are lost. Extending design effort earlier, with the support of digital and robotic tools, could help anticipate complexities before they escalate on site. In addition, recent years reflect that the spread of robotics across manufacturing has steadily reduced the cost of machines. What was once the privilege of large-scale industries is becoming more accessible, opening possibilities for adaptation in specialized fields such as heritage restoration, where accuracy and adaptability are essential. Rising labor costs further support this shift. In heritage contexts, the challenge is amplified, since conservation relies on highly skilled specialists whose work is at a premium and expertise is both rare and expensive. Robotics are not intended to replace this craftsmanship but to complement it, taking on tasks that involve risk, physical intensity, or difficult working conditions, while enabling human expertise to focus on more delicate interventions that demand judgment and cultural understanding.

Robots Ubiquity and Manual Labor Fall

The construction industry has been shaped in recent decades by cycles of crisis and recovery that have redefined the availability, cost, and perception of labor. One of the most consequential turning points came with the 2008 global recession, when US construction employment declined by 1.5 million jobs between 2007 and 2009, the steepest percentage drop of any nonfarm industry supersector (Hadi 2011). The downturn highlighted the sector's volatility and reinforced its image as an insecure and physically demanding field. The reputational damage lingered well after the economy began to stabilize, leaving contractors struggling to attract younger generations into a profession increasingly perceived as dangerous, dull, difficult, and dirty (Keramas 1998).

The global health crisis of 2020 a decade later compounded these pressures. Work stoppages, supply chain disruptions, and heightened safety concerns accelerated attrition from an already shrinking workforce.[1] Older, skilled tradespeople retired earlier than expected, while fewer apprentices entered training pipelines disrupted by school closures and reduced in-person learning opportunities. In many countries, immigration restrictions tightened during the pandemic, further constraining the flow of labor that had historically supported construction and restoration. These losses widened the existing skills gap, particularly in heritage work, where specialized craftsmanship is indispensable but increasingly scarce.

At the same time, labor costs have continued to rise. In the United States and across Europe, the hourly compensation of skilled construction workers has grown steadily over the past two decades,[2] outpacing inflation in many regions. The premium attached to heritage labor is even more pronounced: conservation work relies not only on technical skills but on a body of tacit knowledge built through practice, first-hand experience, material expertise, and craft traditions that are transmitted within professional communities over time. Rising costs reflect both scarcity and necessity, reinforcing the economic pressure on clients and institutions responsible for maintaining historic structures.

Against this backdrop, the role of robotics has become more visible. Historical data from the manufacturing sector reveals a consistent inverse relationship: as labor costs climb, the cost of robots falls (Fig. 7.3). From the 1960s through the 2020s, the average price of industrial robots in the United States dropped by half in inflation-adjusted terms, while their capability, durability and life expectancy more than doubled (Fig. 7.4). By the early 2000s, the operational cost per year of a robot was a fraction of its mid-century equivalent. Average prices fell from over $200,000 in the 1960s to around $112,000 in the 2000s, and toward $70,000 in the 2020s.[3] This trajectory was mirrored in Europe, particularly in leading industrial economies such as Germany and France, where labor costs in the automotive sector rose steadily while automation became more affordable and widespread.

The result was a dramatic increase in robot density across manufacturing. By 2023, there were on average 162 robots per 10,000 employees in factories worldwide, with much higher concentrations in countries like South Korea (1,012), Singapore (770), China (470), Germany (429), and Japan (419). Even in the United States, where adoption was comparatively

Figure 7.3 Declining robot prices compared with rising labor costs in the United States. Index of average robot prices and labor compensation in manufacturing (1990 = 100). Source: McKinsey analysis using data from the Federal Reserve Bank of St. Louis and the International Federation of Robotics, in *Automation and the Talent Challenge in American Manufacturing*, 2024, by Jonathan Tilley, Julian Salguero, and Jacob Achenbach.

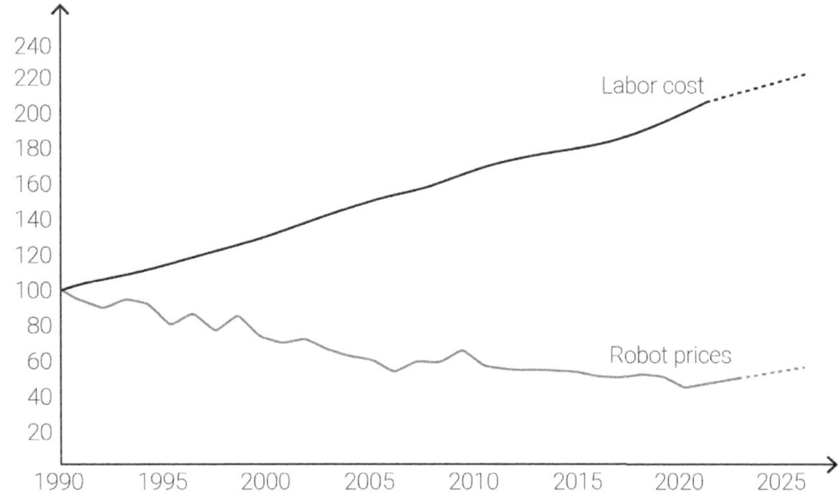

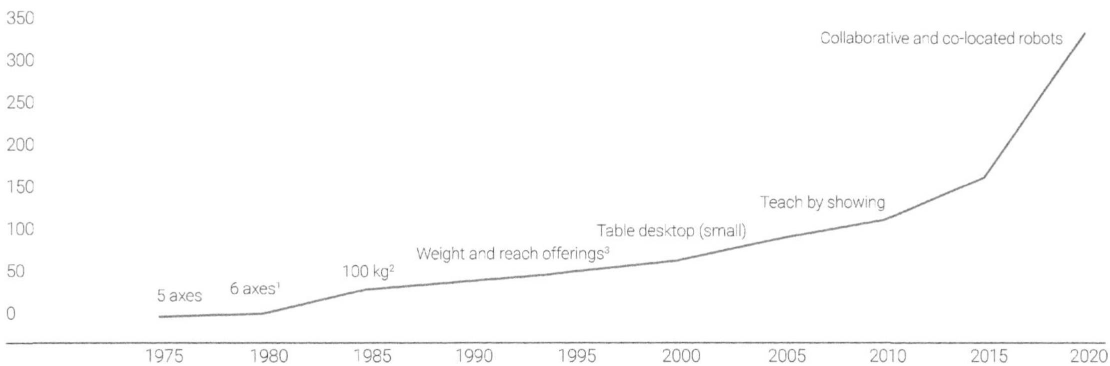

	1975	1980	1985	1990	1995	2000	2005	2010	2015	2020
Number of axis	5	6 to 8	12		21	27			32	
Capacity - kg	6	60	90	>120				1.000		
Max reach, m	1		2					3		

[1] Allows arc welding, adhesives dispensing, machine loading
[2] Spot welding, materials handling.
[3] All application areas; right size for the task.

Figure 7.4 Growth of the robotics market. Expanding variety, size ranges, and capabilities have fueled steady market growth. Source: McKinsey analysis, in *Automation, Robotics, and the Factory of the Future*, 2017, by Jonathan Tilley.

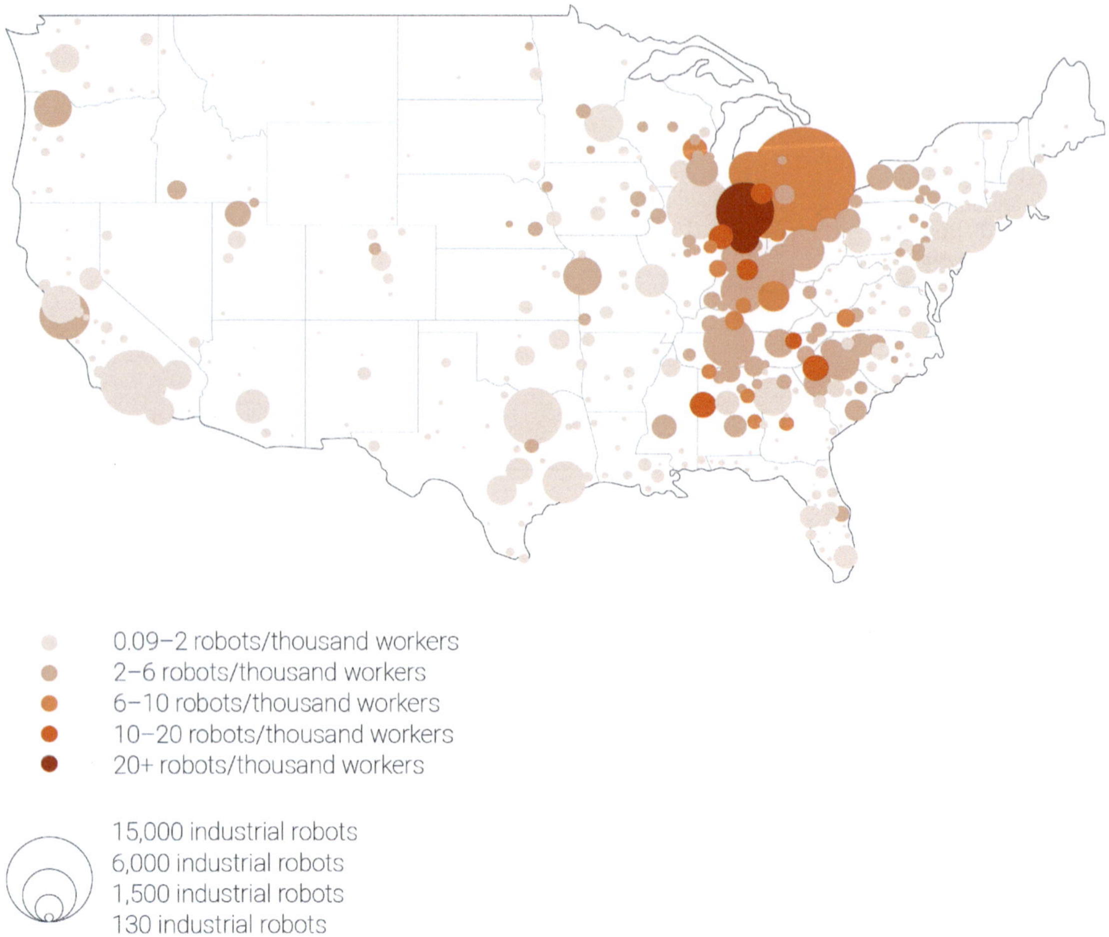

Figure 7.5 Distribution of industrial robots in the United States. As of 2024, approximately 380,000 robots are in operation on factory floors, with Michigan alone accounting for 12 percent of the national total due to its concentration of auto manufacturing. Metro Detroit, as an "auto-intense" region, has three times more robots than the average US metropolitan area. Robots are primarily used for welding, painting, assembly, materials handling, and packaging. Their numbers grew sharply during the post-crisis auto boom between 2010 and 2015. The US map is based on Brookings analysis of International Federation of Robotics data, in Mark Muro, *Where the Robots Are*, 2017.

slower, the ubiquity of robots in manufacturing (Fig. 7.5) reached 295 per 10,000 employees.[4] These figures highlight a long-term realignment: robots became increasingly attractive not only because they could deliver consistent performance, but because their relative cost against human labor tilted ever more in their favor.

For construction and restoration, this dynamic is beginning to exert similar pressure. While robots in building practice have lagged behind those in manufacturing, the same economic logic applies. Construction has long been resistant to full automation due to the variability of sites, materials, and conditions. Yet as robot costs continue to decline and labor shortages persist, the incentive to adapt robotic technologies to unpredictable environments is growing stronger. The construction sector may also benefit from the second life of machines: industrial robots decommissioned from high-volume manufacturing can be repurposed at lower cost for custom applications, including heritage restoration. This recycling of automation assets lowers the barrier to entry and creates new opportunities for experimentation.

European policy frameworks have recognized the urgency of this transformation. Funded programs like the *European Green Deal*, *Horizon 2020* and *Horizon Europe 2021–2027* have explicitly supported research into construction robotics, highlighting both the potential for innovation and the looming workforce challenges. Looking ahead to 2035, an estimated 4 million workers in the European Union will need to be replaced, primarily due to retirements.[5] Alongside this demographic turnover, the adoption of new technologies, including automation and advanced industrialized methods of prefabricated production, is expected to reshape the skills profile of construction work, demanding greater technological literacy and adaptive expertise.

The narrative is not one of replacement but of adaptation: the demand for labor is not disappearing, but the profile of that labor is evolving toward supervisory, planning, and design roles that can work in concert with machines. The future of heritage restoration in particular will depend on this balance. Robotics are not positioned to supplant the craftsmanship, judgment, and cultural understanding that conservation demands. Instead, robotics offers the promise of assuming tasks that are hazardous, physically intensive, or logistically complex: heavy lifting, working in unstable or inaccessible conditions, and handling repetitive processes such as scanning, cleaning, or material deposition. This ultimately allows artisans to focus on interventions where sensitivity to material, context, and history is irreplaceable.

Seen in this light, the fall of labor since 2008 and the disruptions of the 2020 pandemic are not isolated crises but part of a broader reconfiguration of how building work is conceived and performed (Fig. 7.6). Rising labor costs and declining robot costs form two sides of the same equation, reshaping incentives across industries. For heritage restoration, this convergence suggests both risk and opportunity: risk, because the shortage of skilled craftspeople threatens the continuity of cultural knowledge; opportunity, because automation and robotics can be mobilized to safeguard that knowledge by creating new, hybrid practices that combine computational and machine-guided fabrication with human capability.

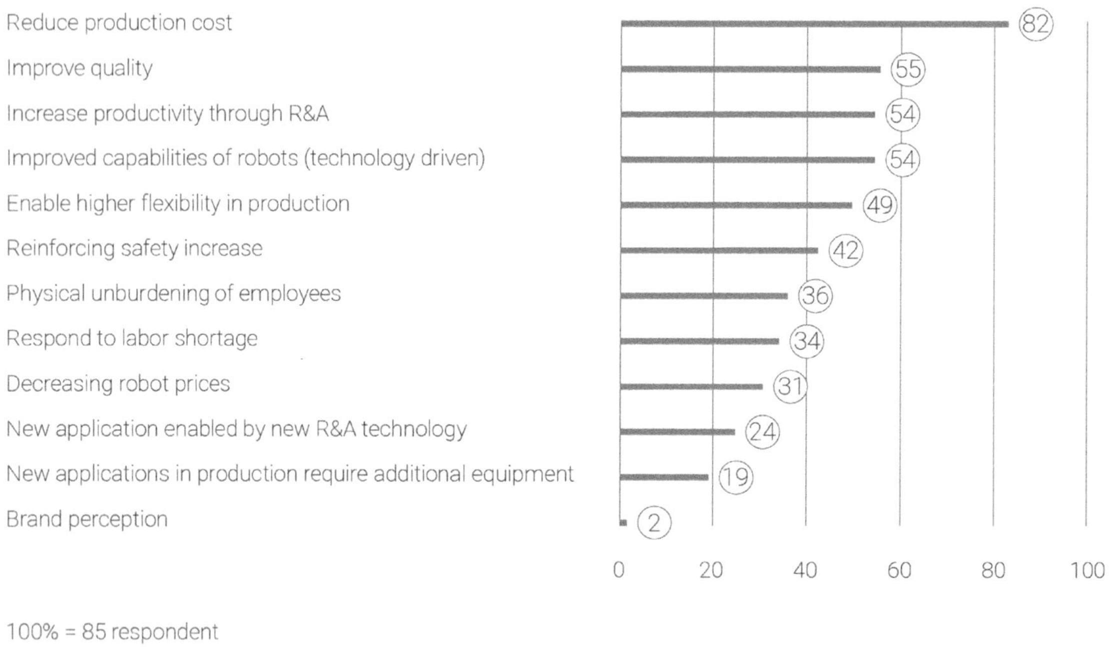

Figure 7.6 Main drivers of investment in robotics and automation solutions. Source: McKinsey *2018 Global Robotics Survey*, published in *Industrial Robotics: Insights into the Sector's Future Growth Dynamics.*

Mapping Initial Applications in Building Renovation

Although robotics in construction is evolving rapidly, cultural heritage restoration will not be the first field to adopt it. Before these technologies are applied to historic buildings, architecture will likely undergo a phase of experimentation, during which confidence is built around new workflows and their capacity to interact with existing structures. A more immediate arena for robotic intervention lies in the reconstruction of the general building stock. Deep renovation, retrofitting, and refurbishment[6] provide opportunities to test constraints, refine operational methods, and directly engage with the material conditions of existing spaces. In this sense, renovation projects represent an advantageous proving ground, setting the optimal conditions for transformative change in the construction sector and encouraging the development of innovative solutions.

The urgency of these applications becomes clear when considering the impact of the built environment, which is estimated to account for 42 percent of annual global CO_2 emissions. Extending the lifespan of buildings through renovation has been shown to be more effective than demolition, as it reduces embodied energy, avoids demolition waste, and lessens the demand for new construction materials. While new buildings may perform better in terms of operational energy, they also require significantly higher embodied energy, making renovation a generally more sustainable strategy. Within frameworks such as the *Moratorium on New Construction*, prioritizing the care of existing heritage buildings positions the discipline to mitigate, repair, and correct the damage produced by contemporary construction practices, rather than perpetuating them (Malterre-Barthes 2025).

Across Europe, the existing building stock illustrates the scope of the task ahead. As of 2024, 35 percent of the EU's building stock is more than 50 years old, and 80 percent was built before the 1990s. These buildings present significant performance and efficiency shortcomings, demanding widespread renovation and retrofitting to meet climate and energy targets.[7] By 2050, new construction is expected to account for

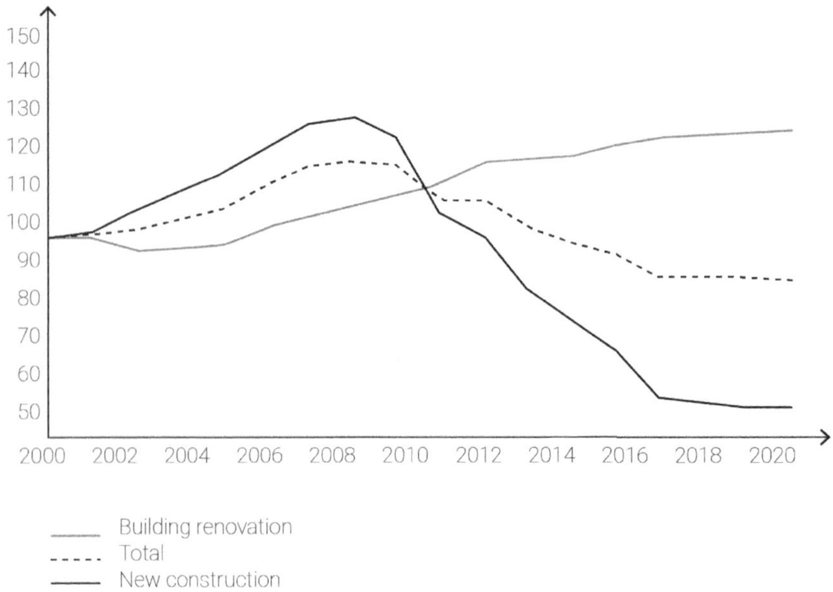

Figure 7.7
Investments in residential construction in Italy (index: 2000 = 100). Diagram based on ANCE data. Over the last decade, the Italian construction industry has been marked by the 2007 building crisis, the aging of existing building stock, and a slowing national economy, which together emphasized the need for greater investment in renovation.

only 10–25 percent of the total building stock, confirming that the majority of architectural work in the coming decades will involve upgrading what already exists. Italy reflects a similar situation, where renovation investments first surpassed those in new construction about a decade ago (Fig. 7.7), a shift tied to the fact that over half of the residential stock predates 1970, an era marked by post-war reconstruction, the economic boom, and large-scale affordable housing programs. While these initiatives expanded housing availability, they were carried out with limited attention to energy efficiency, seismic safety, accessibility, or comfort. As a result, much of the building stock continues to suffer from structural weaknesses, high energy use, elevated management costs, and adverse environmental impacts.

As a result, there is an opportunity for maintenance, care, and renovation to occupy a more prominent position within the architectural profession, not as marginal tasks but as central disciplinary concerns (Sample 2016). These contexts also present the possibility for robotics to be tested, adapted, and refined before extending their application into the more sensitive realm of cultural heritage.

Initial attempts to apply robotics in renovation already suggest a range of possibilities. The *Hephaestus* project,[8] for example, developed a prototype cable robot designed to build, repair, and maintain building façades, alongside a curtain wall system suitable for robotic assembly and a business plan aimed at commercial adoption. Similarly, the *P2-Endure* project[9] explores deep renovation through prefabricated plug-and-play systems combined with on-site robotic 3D printing, offering non-invasive and reversible solutions that allow for aggregation and customization in response to complex design requirements. Such projects demonstrate how robotics can be positioned within the broader field of redevelopment and retrofitting, not only to assist industry practice but also to engage with the geometric, technical, and organizational complexities inherent in constructed buildings. As research advances, predictive models are expected to further reduce design errors and enable more rigorous testing, gradually strengthening confidence in robotic applications for existing structures.

The transition from research to real-world application is already visible in construction practices that integrate robotics and circular design strategies, with examples of robotic systems capable of recycling construction materials for reuse in new projects. These developments point toward broader shifts in construction culture, where automation and sustainability intersect.

One of the most comprehensive steps in this direction can be seen in the work of the Danish architecture studio Lendager, which has made radical design decisions to show how reuse, recycling, and upcycling can operate in a successful dialogue with architectural quality. Their *Resource Rows* project in Copenhagen's Ørestad district is recognized as the world's first residential development constructed almost entirely from materials sourced from demolished buildings (Fig. 7.8).

The project reintroduces value into discarded material streams: upcycled bricks, waste wood, and even a recycled concrete beam repurposed as a pedestrian bridge. Old windows and other salvaged materials were also incorporated, reducing the project's overall environmental impact. A central strategy was the reuse of existing façade systems.

Figure 7.8 *Resource Rows* in the Ørestad district of Copenhagen, designed by Lendager, recognized as the world's first residential project constructed almost entirely from salvaged materials sourced from demolished buildings. Photo by Rasmus Hjortshøj.

Cement-mortar brick facades, cut from abandoned houses in one-by-one meter sections, roughly 3.3×3.3 feet, were reassembled into new modular façade elements (Fig. 7.9). This method produced a patchwork aesthetic (Fig. 7.10), with a practical reasoning. Unlike traditional lime mortar, the cement mortar widely used in Denmark since the 1960s is harder than the bricks themselves, making individual brick reuse impossible. Sawing out entire wall sections was therefore the only feasible way to reclaim these materials, and this process in turn generated new building elements in the form of "super-bricks." The large modules of salvaged masonry could be stacked, shifted, and recomposed into new façade assemblies.

Figure 7.9 Cutting process of modular façade elements, sawn from existing buildings for reuse in new construction. Photo by Rasmus Hjortshøj.

Each reused brick is estimated to have prevented about 500 grams of CO_2 emissions, and when scaled across the project, this approach reduced the carbon footprint of construction by approximately 12 percent compared to conventional building practices. At the same time, the modular character of the upcycled bricks offered opportunities to test automation in façade construction, with robotic systems, single-task façade bots, deployed to explore how scripting and automation could streamline the

Figure 7.10 Façade pattern made of large salvaged masonry modules. Photo by Rasmus Hjortshøj.

dismantling and reconstruction of building skins (Fig. 7.11). Therefore, *Resource Rows* functioned both as a demonstration of circular building practice and also as a prototype for integrating robotic fabrication into large-scale material reuse.

Whether in new construction, deep renovation, or heritage projects, the long-term promise of the Fourth Industrial Revolution lies in bringing these processes into a single integrated digital infrastructure, where the construction industry can position itself fully within the broader ecosystem of digital transformation. For this to occur, informational tools must be capable of supporting complex decision-making, connecting diverse skills across scales and across the entire life cycle of buildings. The path forward is gradual: advanced technologies must first be tested and refined in renovation and new construction before extending into the more sensitive terrain of cultural heritage. In this way, architectural practice builds both the technical capacity and the cultural confidence required to responsibly apply digital tools to the conservation of irreplaceable assets.

Figure 7.11 Single-task façade bots tested to automate the recovery of large masonry modules. Photo by Lendager.

Education and New Professions

The future of design and construction points to the re-emergence of the master-builder, a generalist professional (Epstein 2019) capable of orchestrating the entire production chain. This role is not a nostalgic return to pre-modern practice, but rather a new figure who synthesizes the complexities of Industry 4.0, Work 4.0, and flexible Construction 4.0. The future master-builder will need to understand and lead workflows that are aligned with the logic of advanced manufacturing processes, where architecture and industry converge. In the context of restoration, this generalist role becomes even more critical, since heritage projects demand the ability to interpret historical knowledge while deploying contemporary tools for design, fabrication, and site work.

Education is set for success if it evolves to prepare such professionals (Doepker 2019). The integration of computation and digital fabrication into architectural pedagogy has already begun to reshape how design is taught and practiced. Experimental approaches in academia—robotics labs, computational design studios, and fabrication research—are not isolated curiosities. Instead, they provide the groundwork for innovation that can influence professional practice and construction at large. These investigations connect with architectural-executive practice, embedding industrial processes and technical know-how into academic training so that students can transition from concept to built form with greater fluency.

Recent advances in robotic research highlight this trajectory.[10] The study of robotic applications in architecture has deepened the dialogue between academic research, material science, and the construction sector. Universities and research centers are increasingly involved in on-site prototyping of building elements, bridging laboratory experimentation with real-world construction challenges. Transitioning from prototype workflows to full-scale construction systems will require coordinated efforts to address sector-specific challenges and advancements. These advances will accelerate technology transfer and encourage investment, creating a feedback loop that strengthens both academic inquiry and industry application.

To support this transformation, education needs to broaden its scope. Future professionals will not only need technical expertise but also the ability to navigate a landscape shaped by smart technologies such as AI, AR/VR, big data, automation, and the Internet of Things. Preparing students for this environment requires a dual approach: strengthening STEM (Science, Technology, Engineering, Mathematics) education while also cultivating creativity, imagination, and social intelligence (Neumeier 2013). Expanding access to STEM learning is particularly important for addressing the persistent gender gap in construction, where building site operations have historically been dominated by male voices. The future of construction will depend on new perspectives and broader participation, ensuring that women and underrepresented groups contribute to reshaping site practices.

Equally important is cultivating curiosity, problem-solving, and resilience through environments that support experimentation and iterative failure. Students should be encouraged to develop skills for careers that may not yet exist. According to the World Economic Forum's *Future of Jobs Report 2025*, nearly two-fifths (39 percent) of current skill sets are expected to be transformed or become obsolete between 2025 and 2030 (World Economic Forum 2025). This demonstrates the need for education to extend beyond early training into systems of lifelong learning and career adaptation. Personalized teaching through accessible technologies such as AI and machine learning can help achieve this, enabling diverse learning paths and preparing learners to remain versatile across changing roles. Schools should also provide spaces where physical and digital tools can be explored side by side, reinforcing the "phygital" nature of future practice. At the same time, students should be trained to think globally (Penprase 2018), engaging with the demands of international contexts and emerging markets. Stronger ties between higher education and industry will be essential, with post-secondary institutions serving as bridges that plant seeds of continuous learning and expand opportunities beyond the traditional curriculum.

Designers of the future will need new skills and adaptable approaches to practice (Deutsch 2015). Process information can serve as the connective tissue linking designers, contractors, owners, and occupants into data-enabled project teams. Education therefore carries the responsibility of shaping the next generation of master-builders: professionals who are not only designers but also curators of processes, materials, and collaborations. More voices are needed in this ongoing conversation, but it is clear that higher education plays a decisive role in preparing new professionals for the Fourth Industrial Revolution.

Coda

Architecture has long questioned what culture, tradition, and monumentality mean. In *Architecture, You and Me*, Sigfried Giedion, Josep Sert, and Fernand Léger outlined nine points of monumentality: the expression of collective will, enduring value, synthesis of function and form, unity of the arts, scale and proportion, context and setting, construction and materiality, tradition and innovation, and emotional impact. Together, these qualities remind us that monuments are not defined by size or material alone, but by their ability to embody shared values and transmit memory across generations.

Preserving such testaments of human culture has always been a priority, but today it intersects with a new horizon. *Robots for Restoration* suggests that monumentality can be reimagined through intelligent machines and digital workflows, extending the dynamics of embodied culture. Each of the nine points gains fresh relevance: collective will in the ethics of preservation, enduring value in long-term data and material care, synthesis in the blending of human and robotic labor, unity of the arts in hybrid phygital forms, proportion and setting in digitally enabled reconstructions, material authenticity through advanced fabrication, tradition renewed by innovation, and emotional impact sustained as heritage enters algorithmic and automated domains.

Every period creates its own symbols of memory. Our age will not only preserve stone and mortar, but also craft new monuments through code, data, and robotic engagement with the built world. The task ahead is to ensure that these tools remain cultural instruments, capable of carrying tradition forward while inspiring awe and continuity. Restoration, in this sense, is not simply about the past, it is a way of designing the future of memory itself.

Notes

1 In Europe, the post-2020 "great resignation," a wave of voluntary job departures and workforce exits that followed the global health crisis, led to the loss of around 435,000 construction jobs, and by 2022, employment still remained 50,000 workers below pre-pandemic levels. Data from the European Centre for the Development of Vocational Training (Cedefop), *Construction Workers: Skills Opportunities and Challenges, 2023 Update.*

2 U.S. Bureau of Labor Statistics, *Databases*, https://www.bls.gov/; Eurostat, *Database*, https://ec.europa.eu/eurostat/.

3 The data on the declining cost of industrial robots in the United States between the 1960s and the 2000s is drawn from *Robot Technology Fundamentals*, by J. G. Keramas, pages 3–7. The analysis demonstrates that while purchase prices dropped significantly in inflation-adjusted terms, the durability and life expectancy of robots increased, lowering the annualized cost of operation. Data for the 2020s has been added from Statista, "Average cost of industrial robots in selected years from 2005 to 2017 with a forecast for 2025 (in U.S. dollars)," which shows that this long-term downward trend in robot pricing continues into the present.

4 Data on global robot density is from the International Federation of Robotics, *World Robotics Report 2024*. The report shows that in 2023, the world average reached 162 robots per 10,000 employees in manufacturing, with regional averages of 219 in Europe, 197 in North America, and 182 in Asia.

5 Data from the European Centre for the Development of Vocational Training (Cedefop), *Construction Workers: Skills Opportunities and Challenges,* 2023 Update.

6 Deep renovation denotes comprehensive upgrades that restructure a building to reduce energy consumption by at least 60 percent compared to pre-intervention levels. Retrofitting adds systems or technologies to improve efficiency, while refurbishment focuses on repair and upgrading to restore functionality without the extensive performance gains of deep renovation.

7 The European Commission's long-term strategy aims for climate neutrality by 2050, achieving net-zero greenhouse gas emissions in line with the Paris Agreement. This target, enshrined in the European Climate Law, forms a central pillar of the European Green Deal.

8 The Hephaestus (Highly automatEd PHysical Achievements and PerformancES using cable roboTs Unique Systems) project, funded by the EU's Horizon 2020 program, developed a cable-driven robotic system for automating façade installation, maintenance, and repair. It includes a modular end-effector capable of performing tasks such as scanning, prefabricated panel installation, painting, cleaning, and minor repairs, while improving safety, quality, and efficiency in high-rise construction scenarios.

9 The P2-Endure (Plug-and-Play product and process innovation for Energy-efficient building deep renovation) project, funded under Horizon 2020, focused on deep renovation strategies, combining prefabrication, BIM-based workflows, and robotic processes. Demonstration projects across Europe highlighted flexible, scalable, and reversible approaches to upgrading existing buildings.

10 To understand the future of technology applications, it is useful to refer to research roadmaps that outline the anticipated development of a technology and the steps toward its effective implementation. One example is the *Canadian Construction Automation and Robotics Roadmap—Report 2025*, which sets out a 15-year vision for how robotics, AI, and digital tools can reshape construction by improving productivity, safety, and sustainability, while also addressing labor shortages and highlighting barriers such as fragmented adoption, limited R&D investment, and the need for shared standards and training.

References

Deutsch, Randy. 2015. *Data-Driven Design and Construction: 25 Strategies for Capturing, Analyzing and Applying Building Data*. John Wiley & Sons.

Doepker, Megan. 2019. "5 Ways Students Can Graduate Fully Qualified for the Fourth Industrial Revolution." *World Economic Forum*, January.

Epstein, David. 2019. *Range: Why Generalists Triumph in a Specialized World*. Riverhead Books.

Hadi, Adam. 2011. "Construction Employment Peaks before the Recession and Falls Sharply throughout It." *Monthly Labor Review* 134: 24.

Keramas, James G. 1998. *Robot Technology Fundamentals*. Delmar Publishers.

Malterre-Barthes, Anne-Charlotte. 2025. *A Moratorium on New Construction*. Critical Spatial Practice. Sternberg Press.

Neumeier, Marty. 2013. *Metaskills: Five Talents for the Robotic Age*. New Riders.

Penprase, Bryan Edward. 2018. "The Fourth Industrial Revolution and Higher Education." *Higher Education in the Era of the Fourth Industrial Revolution* 10 (1): 978–81.

Sample, Hilary. 2016. *Maintenance Architecture*. The MIT Press.

World Economic Forum. 2025. *The Future of Jobs Report 2025*. World Economic Forum.

INDEX

Note: Page numbers in *italics* refer to figures and those with "n" refer to notes.

For Product Safety Concerns and Information please contact our
EU representative GPSR@taylorandfrancis.com Taylor & Francis
Verlag GmbH, Kaufingerstraße 24, 80331 München, Germany